Manual Series - I

Molecular Biology & Biochemistry — A Lab Manual

NIPA® GENX ELECTRONIC RESOURCES & SOLUTIONS P. LTD.
New Delhi-110 034

About the Authors

Dr. H.P. Puttaraju, M.Sc., Ph.D., FRES is a Professor of Sericulture/ Life Sciences and Co-ordinator, Biological Sciences (Integrated M.Sc.) of Bangalore University, Bangalore, Karnataka. With acquiring advanced training at Cardiff University of Wales, UK, he has investigated and studied the importance of Wolbachia endosymbiont in insect and insect pests using molecular techniques. Using DNA markers he and his group have analyzed several plant and insect genotypes. So far he has successfully guided 11 Ph.D and 7 M. Phil students on various aspects of Life Sciences. He has authored/edited 2 books in his field of interest and published over 70 research papers in reputed national and international journals. He is recipient of Sir C.V. Raman Young Scientist Karnataka State Award, Young Histochemist Award, and is a Fellow of Royal Entomological Society, Fellow of Commonwealth Academic Staff Scholarship Commission, UK visiting fellow under INSA-Royal Society bilateral exchange programme.

Dr. B.M. Prakash, M.Sc., Ph.D. has completed his doctoral studies with Seri-Biotechnology as the field specialization at Bangalore University and joined JNCASR for post doctoral research. His research in Ph.D programme includes understanding the infection frequency of Wolbachia (vertically transmitted bacterial endosymbiont) in insect pests associated with sericulture and its possible role in controlling uzifly, *Exorista sorbillans*, a serious tachinid endoparasitoid of silkworm, *Bombyx mori* L.

Mr. B.C. Keshava Murthy, M.Sc., Ph.D. has completed in Ph.D. degree very recently with specialization in Plant Biotechnology. He joined MAS Lab. Department of Genetics and Plant Breeding, University of Agricultural Sciences, Bangalore for Post Doctoral Research. His area of interest lies in application of DNA markers in Plants with particular references to Molecular Breeding and Genome Mapping.

Molecular Biology
&
Biochemistry

—A LAB MANUAL—

Authors

H. P. PUTTARAJU
Professor, Sericulture/Life Sciences and Co-ordinator
Biological Sciences (Integrated M.Sc.)
Bangalore University, Bangalore, Karnataka

B. M. PRAKASH
Department of Sericulture/Life Sciences
Bangalore University, Bangalore, Karnataka

B. C. KESHAVA MURTHY
Department of Genetics and Plant Breeding
University of Agricultural Sciences
GKVK, Bangalore, Karnataka

NIPA® GENX ELECTRONIC RESOURCES & SOLUTIONS P. LTD.
New Delhi-110 034

NIPA® GENX ELECTRONIC
RESOURCES & SOLUTIONS P. LTD.

101,103, Vikas Surya Plaza, CU Block
L.S.C. Market, Pitam Pura, New Delhi-110 034
Ph : +91 11 27341616, 27341717, 27341718
E-mail: newindiapublishingagency@gmail.com
web: www.nipabooks.com

For customer assistance, please contact
Phone: + 91-11-27 34 17 17
Fax: + 91-11- 27 34 16 16
E-Mail: feedbacks@nipabooks.com

ISBN: 978-81-19235-25-4

Composed and Designed by NIPA®.

Preface

Advances in Biotechnological tools have helped in more unified understanding of many plant and animal genomes and most of the experiments which were unimaginable in earlier years have been performed very efficiently. Most of the experiments involved in molecular biology are difficult to analyze, understand and perform for the beginners. The basic information about techniques in molecular biology are covered in many introductory molecular biology and biotechnology books but protocols of such experiments have been mentioned in very few book chapters. Finding these protocols is very cumbersome for students and researchers who are at the beginning stages of their experiment/research and theoretically any student can learn about any techniques but performing the same at laboratory level is very difficult without first hand experience. Just by learning protocols one cannot do molecular biology experiments and the principle behind becomes very crucial for the effective solution to a particular biological problem.

This lab manual gives information on the protocols, principles and applications behind mostly performed experiments and it has been written keeping in mind the requirements of graduate/postgraduate students and research scholars. With the boom in biotechnology research in India many new molecular biology laboratories have been established in many Universities and Research institutes. There has been steep increase in the number of students studying molecular biology and biotechnology courses all over India. The present book chapters contain information on basic protocols in simplified manner which is very easy to learn and perform. This book further gives first hands-on information and it has been divided into 10 chapters with each chapter containing introduction, principle, protocol, applications and troubleshooting.

Manuscript of the book contains many protocols and figures which were collected from various websites, research papers and training manuals, we acknowledge all those individuals involved directly or indirectly in preparation of this Laboratory Manual. Most of the existing protocols have been mentioned in a simplified manner just to aid easy understanding for students and helping the same to perform very effectively.

June, 2008

H.P. Puttaraju
B.M. Prakash
B.C. Keshava Murthy

Introduction

Molecular biology deals with the study of structure and function of molecules mainly biological molecules. Nucleic acids are the most important biological molecules.

Nucleic acids

Nucleic acids are macromolecules present in all the living cells. There are two types of nucleic acids, Deoxyribo nucleic acids (DNA) and Ribosenucleic acid (RNA). The former is mainly found in nucleus and the latter in the cytoplasm of the cell. Nucleic acids were first discovered while investigating the action of pepsin on pus cells by Frederick Miesher in the year 1868. The undigested residue compound obtained from the nuclei of pus cells was named by him as Nuclein. Since it was acidic in nature, Altmann in 1899 called it as Nucleic acid. Later the occurrence of nucleic acid was obtained in the cell of both plants and animals. Further, two type of Purines, Adenine and Guanine; two types of Pyrimidines, Thymine and cytosine; 5-carbon ribose sugar and phosphoric acid were identified in the nucleic acids. The two types of nucleic acids, DNA and RNA were recognized on the basis of the types of pentose sugars. In 1953, James D Watson and F.H.C. Crick constructed the double helix model of DNA based on its physiochemical nature and in 1967, Korenberg synthesized the DNA in the laboratory. This discovery led to our understanding of the structure, replication and transmission of genetic information.

Deoxyribose Nucleic acid (DNA)

DNA is the macromolecule of high molecular weight at the order of millions of daltons. It is the master molecule of biological system and form the genetic material. The DNA is present in almost all the organism excluding a few prokaryotes like plant viruses which contain RNA. The DNA is normally found inside the nuclei of the cells. The chromosomes found in the nucleus are made up of DNA and other associated proteins to form nucleoproteins. Among the prokaryotes like bacteria and blue green algae, naked DNA associated with carbon proteins is suspended in the cytoplasm. It is also found in the organelles like mitochondria and chloroplasts of the cytoplasm. Such DNA are called extra nuclear DNA and it brings about cytoplasmic inheritance.

Ribose Nucleic acid (RNA)

RNA is another class of nucleic acid found in living systems. More than 90% of RNA is found in the cytoplasm and the rest in the nucleus. RNA is a macromolecule with a high molecular weight of more than 25,000 daltons to a few million daltons. It is a smaller molecule when compared to DNA. Further the RNA is single stranded unlike DNA and it is a poly nucleotide chain with 5, phosphate and 3, -OH polarity. The nucleotides are made up of ribose sugars (Pentose sugars), Phosphate groups and nitrogenous bases-Adenine, Guanine, Uracil and Cytosine.

The tools and techniques involved in studying these Molecules are presented in this manual.

Content

Chapter 1

Laboratory Equipments and Its Usage

Molecular biology labo-ratories are becoming very common in every University and research institutes and a good laboratory manners is extremely important as many of the equipments are very expensive and sensitive. Hence familiarity with equipments becomes very crucial before starting any experiment.

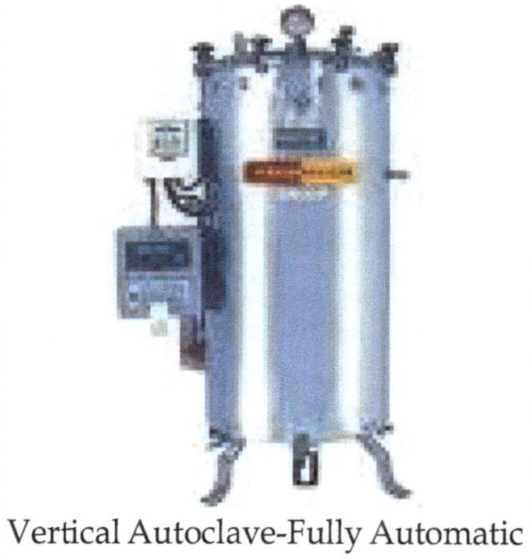

Vertical Autoclave-Fully Automatic

1. Autoclave–for sterilizing solutions, media and other lab-ware.

1a. Ultra low temperature freezer (-80°C) - for long-term storage of chemicals, primers, DNA, etc.

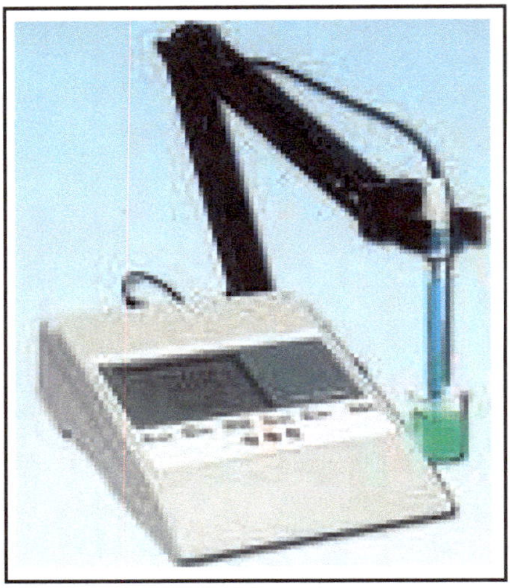

2. pH meter - for adjusting or measuring pH of the solutions and reagents.

3. Micropipettes come in a range of sizes. They have disposable tips that hold tiny amounts of required reagents.

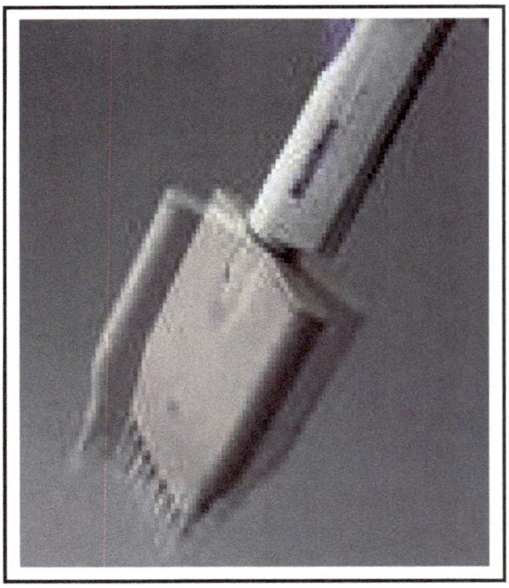

4. Multi channel micropipette for rapid loading of DNA samples and rapid transferring various enzymes/chemicals.

5. Hot plate stirrer-for stirring/heating solutions.

6. Balance - for weighing chemicals needed to prepare solutions or reagents.

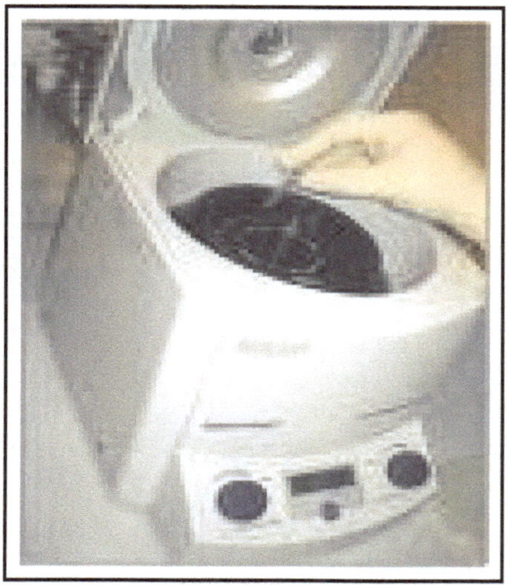

7. Microfuge-spinning & precipitating small amounts of nucleic acids & protein samples at high speed (10,000 - 14,000 rpm).

8. Cold centrifuge for spinning & precipitating of samples at low temperature (1,000 to 18,000 rpm).

9. Incubator – for incubating cultures/ DNA samples for digestion at a specified temperature.

10. Laminar air flow for the preparation of microbial cultures.

11. Fume hood–for preparing or mixing chemicals like acids and others, emitting strong fumes.

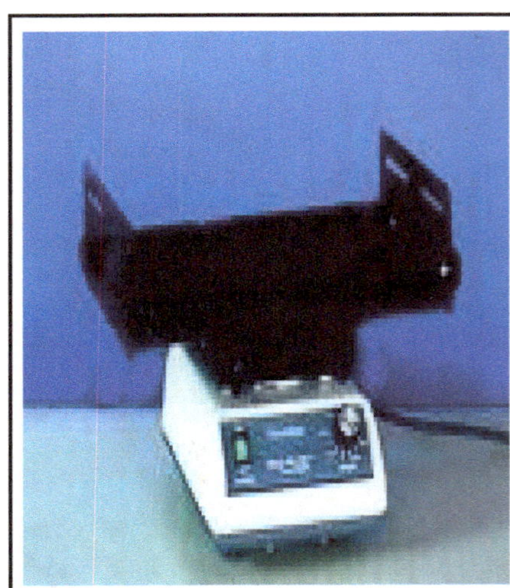

12. Orbital shaker – for shaking samples or various other purposes.

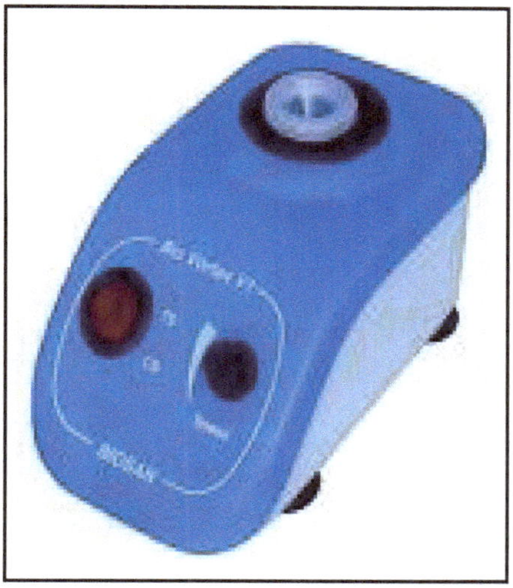

13. Vortex for through mixing of ingradients in vials.

14. Speed Vac- for drying and evaporating liquid from DNA samples in suspension.

15. Microwave oven - for heating solutions or melting agarose gel.

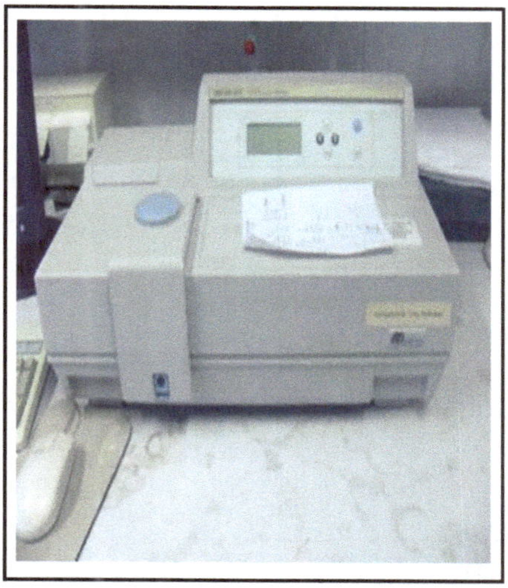

16. Spectrophotometer - Is used to measure the intensity DNA, RNA, Proteins etc.

17. Thermocycler-For amplifying DNA samples using polymerase chain reaction (PCR), denaturation of DNA samples, etc.

18. Gel documentation unit for the documentation of gels and for further analysis of DNA/RNA bands.

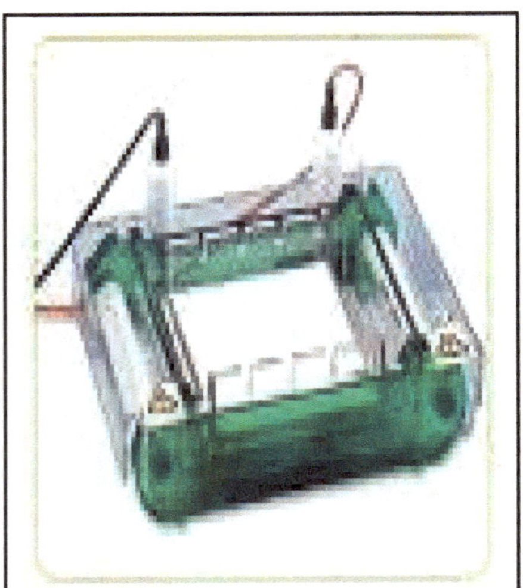

19. Submarine/horizontal Gel Electrophoresis system - for separating fragments of DNA on the agarose gel after PCR amplification and/or digestion

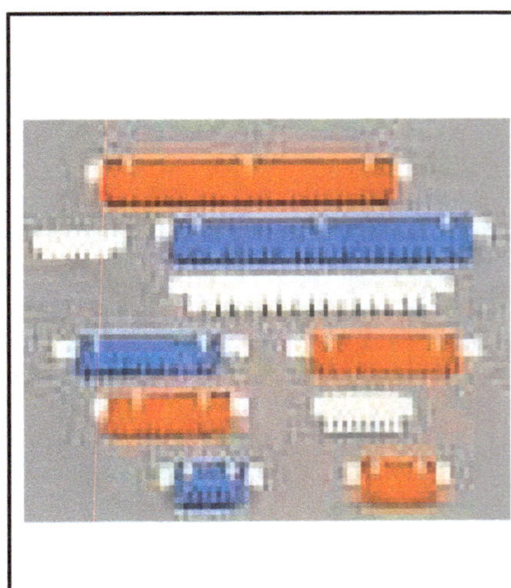

20. Plastic combs used with agarose gels for making different sized wells to DNA samples

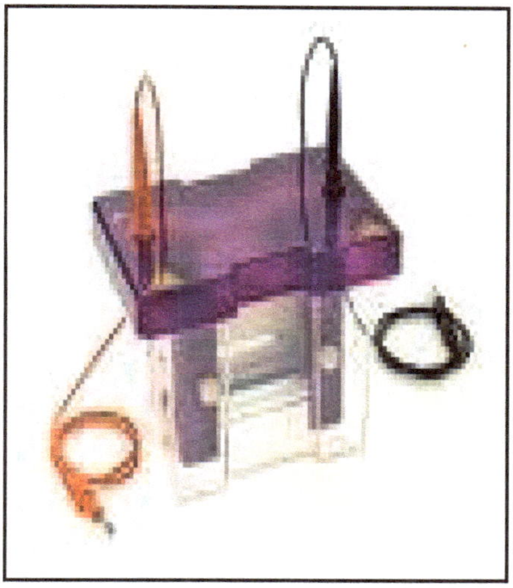

21. Vertical Gel Electrophoresis system - for separating DNA fragments generated by PCR on the denaturing sequencing gel (PAGE system) and for protein separation.

22. UV transilluminator - for visualizing DNA and RNA samples in the agarose gels and polyacylamide gels.

23. Automated sequencer attached to a computer.

❑❑❑

Chapter 2

Buffers

Introduction

This section of the manual contains instructions for preparing stock solutions that are commonly used in laboratory protocols for the analysis of DNA, RNA and proteins and visualization of gene or protein expression. Each of these stock solutions is provided as either the most convenient concentration or the most frequently used stock concentrations to be used in the protocols of an individual laboratory.

All chemicals must be reagent grade or molecular biology grade, and the water used in the preparation of all solutions must be the highest quality available. Use sterile-glass distilled deionised water whenever possible. Unless otherwise stated, most solutions required sterilization either by filtration or autoclaving.

Principle

Biological Buffers

A buffer, as defined by Van Slyke is "a substance which by its presence in solution increases the amount of acid or alkali that must be added to cause unit change in pH". Buffers are thus very important components in experiments designed to study biological reactions by maintaining a constant concentration of hydrogen ions within the physiological range. The pH of mammalian blood is maintained close to 7.38 by buffer systems such as many organic acids, organic bases and proteins.

$$H_2PO_4^- \quad <=> \quad HPO_4^{2-}$$

$$CO_2 \quad <=> \quad H_2CO_3$$

$$H_2CO_3 \quad <=> \quad HCO_3^-$$

In living plants, the normal range of pH in tissues is about 4.0-6.2. It is not as narrowly defined as in mammalian tissues.

A solution containing a mixture of weak acids (HA) and its conjugated base (A^-) that is capable of resisting substantial change in pH on addition of small amount of acidic or basic substances are termed as buffers (see below). Addition of acids to a buffers leads to conversion of some A^- to the HA form; addition of base leads to conversion of some of the HA to the A^- form. As a result, addition of either acid or base leads to change in the A^-/HA ratio. In the Henderson-Hasselbalch equation, which applies to buffer system, the A^- / HA ratio lead to only minor changes in pH within the working range of the buffer.

Henderson-Hasselbalch Equation

The ionization of a weak acid (HA) is termed as Henderson-Hasselbalch equation.

$$HA = H^+ + A^-$$

where HA is a weak acids (Proton donor, conjugated acid) and A is a weak base (proton acceptor, conjugated base). The two classes HA and A are said to form a conjugated acid-base couple. The dissociation or ionization (equilibrium) constant for this reaction K or K_{eq} as follows

$$K \text{ or } K_{eq} = [H^+]\,[A^-] \,/\, [HA]$$

Where parenthesis indicate molar concentration (moles per liter). By taking logarithms and rearranging, the Henderson-Hasselbalch equation is obtained

$$pH = pK + \log [A^-] / [HA]$$

where pH is equal to the negative logarithm of the hydrogen ion concentration ($-\log [H^+]$) and $pK = -\log K$.

$$\begin{aligned} pH &= -\log [H^+] \\ &= \log 1/[H^+] \end{aligned}$$

The definition by which Arrhenius attributed was chosen in order to obtain small positive numbers for the hydrogen ion concentration.

Measurement of pH

The pH can be measured in many ways. An accurate and practical method for measuring pH involves the use of a pH meter. The pH meter is a potentiometer, which measures the potential developed between a glass electrodes and a reference electrode. In modern instruments, the two electrodes are often combined into one electrode, known as a combination electrode. The glass electrode consists of a glass bulb made of very thin, special glass that is permeable to hydrogen ions. The potential is linearly related to the pH. Standardization against a buffer of known H^+ concentration is required since the concentration of H^+ inside the bulb of the glass electrode changes with time. Adjustment for temperature is necessary since the relationship between measured potential and pH is temperature dependent.

The pH scale usually runs from 0 to 14, with 7 representing neutrality and pH value above 7 characterize basic conditions and pH value below 7 represents acidic condition.

Universally applicable buffers for biochemistry and molecular biology must display

- Water solubility
- No interference with biological processes

- Known complex-forming tendency with metal ions
- Non-toxicity
- No interference with biological membranes (penetration, solubilization, adsorption on surface etc.)
- Very low U.V. absorption at wavelength >260 nm

Protocol

Reagents

Preparation and dilution of solution

Percent solution

A solution in which the exact concentration of the solute in 100 ml of a liquid is known as percentage solution. The concentration may be expressed by weight or volume.

Percentage (v/v) = x ml in 100 ml of solution

Percentage (w/v) = x g in 100 ml solution

For example

For percentage

The concentration of commercially available formaldehyde varies from 35 to 40% (v/v). If we want 2% (v/v) solution of formaldehyde, is not correct to take 2 ml of 35-40% (v/v) formaldehyde and add it to 98 ml of distilled water. An acceptable and correct formula for determining the amount of 35% (v/v) formaldehyde to be taken up 100 ml of 2% (v/v) aqueous solution of formaldehyde is as follows.

$$V_1 C_1 = V_2 C_2$$

where

V_1 = initial volume; C_1 = initial concentration

V_2 = final volume; C_2 = final concentration

x ml X 35% = 2% X 100ml

x ml = 2% x 100 ml / 35%

x = 5.71 ml of 35% (v/v) formaldehyde in 94.29 ml of distilled water

For weight

For a 2% (w/v) aqueous solution of potassium chloride, weight out 2 g of KCl and add to 100 ml of distilled water.

The percentage volume of solution can also be prepared by using above formula.

Preparation of concentrated solutions

1. *Molar solution (M) (mole):* The SI unit of amount of substances, equal to the quantity containing as many elementary units as there are atoms in 0.012kg of Carbon-12 or One gram molecular weight of the solute dissolved in 1000 ml of solvent.
2. *Normal solution (N):* one gram equivalent weight of the solute dissolved in 1000 ml of distilled water.
3. *Molal (mol):* containing one mole of solute per kilogram of solvent.

Applications

Buffer solutions are necessary to keep the right pH for enzymes in many organisms to work. A lot of enzymes work only under very precise conditions, if the pH stays too far out of the margin the enzymes slow or stop working and the enzyme can denature, thus permanently disabling its catalytic activity.

A buffer of carbonic acid (H_2CO_3) and bicarbonate (HCO_3^-) is present in blood plasma, to maintain the pH between 7.35 and 7.45.

Industrially, buffer solutions are useful in fermentation processes, and for setting the correct conditions for the dyes used in colouring fabrics. Also, buffer solution can be used in chemical analysis and calibration of pH meters.

Troubleshooting

- Ensure quality of chemical.
- Ensure right quality of water.
- Ensure addition of all the components per buffer.

- Autoclave all the reagents if needed.
- Ensure maintenance of recommended pH.
- Before adjusting pH, ensure that the pH meter is properly working or not.
- Ensure that whether the buffer is light sensitive or not, if the buffer is light sensitive, stored in brown bottle.
- Ensure that the buffer is not older than 3 - 5 months.

References

Bates, R.G. (1973) Determination of pH. Theory and practice, 2nd edn. Wiley, New York.

Blanchard, J.S. (1984) Buffers for enzymes. *Methods Enzymol.*104, 404–414.

Dawson, R.M.C., Elliot, D.C., Elliot, W.H. & Jones, K.M. (1986). Data for Biochemical Research, 3rd ed., Oxford Science Publ.

Ellis, K.J. & Morrison, J.F. (1982) Buffers of constant ionic strength for studying pH-dependent processes. *Methods Enzymol.* 87, 405–426.

Gomori, G. (1955) Preparation of buffers for use in enzyme studies. *Methods Enzymol.* 1, 138-146.

Good, N.E., Winget, G.D., Winter, W., Connolly, T.N., Izawa, S. & Singh, R.M.M. (1966) Hydrogen ion buffers for biological research. *Biochemistry.* 5, 467–477.

Van Slyke, D.D. (1922) On the measurement of buffer values and on the relationship of buffer value to the disassociation constant of the buffer and the concentration and reaction of the buffer solution. *J Biol Chem.* 52, 525–570.

Perrin, D.D. & Dempsey, B. (1974) Buffers for pH and Metal Ion. Control. London: Chapman and Hall.

McLellan, T. (1982) Electrophoresis buffers for polyacrylamide gels at various pH. *Anal Biochem.* 126, 94–99.

❒❒❒

Chapter 3

Cell Culture

Plant cell culture

Introduction

As the human population is expanding there has been increasing concern over whether current rates of food productions can support the growing population. In response to the increasing demand, scientists and farmers have been working hard to come up with new ideas to meet these needs. These ideas range from simple procedures including selection of the best seeds to more complex procedures requiring the transplantation of genes into different plant species. This chapter focuses on one of the important aspects of plant science which is the growth of plant cells in an artificial media. This process is known as plant cell culture and can be applied in many different ways.

Principle

When carrying out a plant cell culture, there are three important things that must be considered. First the plant part of interest must be isolated from the intact plant. Next the appropriate environment to promote optimal growth must be discovered and applied. This may vary depending on the cells of interest. Finally, these procedures must be carried out in a sterile environment to prevent growth of microorganisms. These topics are discussed in addition to applications in which cell culture has been used.

Growth conditions for plant culture must provide it with all essential nutrients which are normally provided by the plant itself. Furthermore, these conditions must be optimized to promote the growth of the particular cell type. The medium, in which the cells are grown, must contain a carbon source, vitamins, salts and other organic supplements. The inorganic nutrients come in different concentrations of macronutrients and micronutrients. Macronutrients include nitrogen, potassium, calcium and magnesium, which are always found in a concentration greater than 0.5mM. The micronutrients include iron, copper, zinc, and cobalt and can be adjusted for maximum growth of each culture type. In addition, medium must provide vitamins which are usually synthesized by plants. In addition to vitamins and nutrients, growth hormones can also influence the growth rate of plants grown in artificial environments. Many plants grown in a conventional manner produce their own growth hormones, however, in culture; artificial hormones are supplied to ensure optimal growth of the plants.

Besides having the appropriate nutrients in the media, the maintenance of sterile conditions is essential for the success of the cell culture allowing it to be free of microorganisms. This requires that all equipment used in creating a cell culture must be sterilized to ensure contamination does not occur. There are many methods of ensuring sterilization with the use of alcohol, flame, and chemicals. Furthermore, containers should be covered at all times to ensure no further airborne contamination.

Culture types

Many plant species can be regenerated in-vitro through several approaches but all require a starting point. This can be anything from a single cell that can be reproduced, a tissue or organ part, or a cut out piece of differentiated tissue (or organ) known as an explant. Once the starting point has been determined, the culture used to grow this part must be considered. There are many different plant cultures all useful for different things. Some of these culture types include the embryo culture, organ culture, callus culture, and cell culture.

The main culture of interest is the cell culture, which allows for the culture of isolated cells from a very small cluster dispersed in a liquid media. This is the culture type that we will be focusing on throughout the remainder of this chapter.

Cell suspension cultures are rapidly dividing homogenous suspensions of cells grown in liquid nutrient media from which samples can be taken. In a cell suspension, a mass of cells, called a callus, must first be collected. The callus can then be suspended in a liquid callus induction media containing all the required nutrients and elements to allow for optimal growth which acts to turn all cells into undifferentiated cells. The cell suspension is then placed on a shaker to allow the cell aggregates to disperse to form smaller clumps and single cells that are equally distributed throughout the liquid media. The cells will continuously grow until one of the factors becomes limiting causing cell growth to slow.

Before this happens, cells are collected and grown on a media containing hormones to activate specific cell growth. For example if the media contains auxin, it may activate formation of adventitious roots, whereas cytokinin would stimulate axillary and adventitious shoot proliferation and regulate differentiation. Once the plants have developed, they can be transplanted into soil where they can continue to grow.

Protocol

Protoplast Preparation – Recipes

I. Protocol for the preparation of metabolically competent protoplasts

Method for isolating large numbers of metabolically competent protoplasts from leaves of monocotyledons (grasses), dicotyledons (such as spinach and sunflower) or from hypocotyl tissue (e.g. *Brassica napus*).

1. Leaf slices of monocots and dicots are prepared by cutting the leaves with a sharp razor blade into segments of 0.6 - 1 mm in size. In the case of dicots the epidermis can be scraped off before cutting by rubbing with fine carborundum powder or with a fine nylon brush.
2. Set up 50 ml of digestion medium (for 10 - 15 g of plant tissue) according to the recipe listed below (Solution I).
3. Incubate the leaf slices in 18 cm-diameter dish containing the digestion medium for 3 hours at 25°C, covered with a plastic film. It may be advantageous to replace the digestion medium at intervals of 1 hour, as the enzymes might become inactivated by substances released from broken cells.
4. After completion of the incubation the digestion medium is carefully removed and discarded. It normally contains very few protoplasts. The plant tissue is then washed 3 times by shaking gently with 20 ml wash medium (Solution II).
5. After each wash the tissue is collected by pouring through a tea strainer (0.6 to 1 mm pore size) and washed for repeated times before filtering through nylon mesh (120 -200 μm pore size) to remove vascular tissue and undigested material.
6. The protoplasts are collected by centrifugation of the combined filtered washes for 4 minutes at 60 - 100 x g and the supernatant is aspirated and discarded.
7. This crude protoplast preparation also contains some cells and chloroplasts and it is important to purify the protoplasts to remove these contaminants. This can be done with solutions of sucrose and sorbitol of different densities.
8. The protoplast pellet is gently resuspended in 45 ml of Solution III and this suspension is divided among two 100 ml centrifuge tubes.
9. To each tube add slowly 6 ml of Solution IV and then overlay this with 5 ml of wash medium (Solution II) to make a three step gradient.

10. Centrifuge at 300 g for 6 minutes.
11. The protoplasts now collect as a band at the interface between the 2 top layers. Carefully remove them with a pasteur pipette.
12. The protoplast should be examined with a light microscope to ensure that the preparation is free of cells and chloroplasts.
13. When a large portion of the protoplasts is pelleted in this sucrose/sorbitol gradient the density of the 2 layers can be increased by adding 5 - 10% Dextran or 10 - 20 % Ficoll to increase the percentage of the floating protoplasts.
14. The purified protoplasts can be concentrated by diluting with 10 ml of Solution II, centrifuge at 100 g for 3 minutes and then resuspend the pellet in a small amount of medium by gently shaking the tubes.
15. The protoplasts are stable for up to 24 h when stored on ice. Photosynthetic activity of the protoplasts can be determined by measuring co-dependent O_2 evolution with an oxygen electrode, provided rapid stirring is avoided as this will break some of the protoplasts. A suitable medium is listed as Solution V.
16. The protoplasts exhibit a relatively broad pH range but at more acidic pH values the bicarbonate concentration should be lowered.

Solutions for Protocol I

Solution I (Digestion medium)		
Composition		**For 50 ml use**
500 mM	D-Sorbitol	4.56 g
1 mM	$CaCl_2$	7.35 mg
5 mM	MES-KOH, pH 5.5	49 mg
2 %	Cellulase Onozuka R10	1 g
0.3 %	Macerozyme R10	0.15 g

The pH must be adjusted to 6.0 with KOH before adding the enzymes.

Solution II (Wash medium)		
Composition		**For 50 ml use**
500 mM	D-Sorbitol	4.56 g
1 mM	$CaCl_2$	7.35 mg
5 mM	MES-KOH, pH 6.0	49 mg

Solution III		
Composition		**For 50 ml use**
500 mM	Sucrose	8.56 g
1 mM	$CaCl_2$	7.35 mg
5 mM	MES-KOH, pH 6.0	49 mg

Solution IV		
Composition		**For 50 ml use**
400 mM	Sucrose	6.80 g
100 mM	D-Sorbitol	0.90 g
1 mM	$CaCl_2$	7.35 mg
5 mM	MES-KOH, pH 6.0	49 mg

Solution V		
Composition		**For 50 ml use**
500 mM	D-Sorbitol	4.56 g
1 mM	$CaCl_2$	7.35 mg
30 mM	Tricine-KOH	538.00 mg
5 mM	D-Sorbitol	0.90 g

Reagents

Calcium chloride dihydrate
Cellulase Onozuka R10
Macerozyme R10
MES (Morpholinoethane sulfonic acid)
Sodium bicarbonate ($NaHCO_3$)
Sorbitol
Sucrose
Tricine (N-Tris (hydroxymethyl) methylglycine)

II. Protocol for the preparation of regeneration competent protoplasts

Method for the isolation of *Brassica napus* protoplasts from mesophyll material derived from one single individual plant (greenhouse grown plants or *in-vitro* cultured shoots) or from young *in-vitro* grown hypocotyls when differed individuals are acceptable.

When using leaf material a dark treatment (24 to 48 h) is recommended for achieving optimal mesophyll protoplast isolation. To obtain sterile hypocotyls, seeds are surface sterilized with 7.5 % calcium hypochlorite for 1 h with continuous shaking, treated with 70 % ethanol for 1 min. and then washed twice in sterile distilled water. The seeds are germinated on MS-medium (Vamling et Glimelius, 1990).

Hypocotyl and mesophyll materials are treated identically for protoplast isolation; in case of modifications for the mesophyll treatment the values are given in brackets.

1. 4 to 5 days (mesophyll 30 days) old Hypocotyls are cut into 0.5 to 1 mm (1 to 2 mm) segments and incubated for 1 h in plasmolyse buffer (Vamling et Glimelius, 1990).

   After plasmolysing the material it is treated with 1 % Cellulase and 0.1 % Mazerocyme in the medium and modified with the addition of 0.4 M sucrose. Hypocotyl material is incubated for 16 to 18 hours at 20 - 22°C in the dark room on a rocking tray with low speed. The mesophyll material is incubated under the same conditions, but without shaking and the time for enzyme treatment can be shortened to 4 - 5 h.

2. The enzyme treated material is filtered through a nylon mesh (50 μm pore size) to separate undegraded material. To the protoplasts containing enzyme solution an equal volume of the salt solution is added.

3. The mixed solutions are centrifuged in a swing-out rotor at 100 x g for 7 min.
4. Protoplasts with intact cell membranes float on the surface and can be removed with a Pasteur pipette.
5. The protoplast suspension is diluted about 10 times in the salt solution.
6. Washing of the protoplasts in the salt solution is repeated once by flotation and replacing the medium.
7. The protoplasts are finally pelleted by centrifugation in culture medium & cultured at a density of $2.5 - 5.0 \times 10^4$ /ml in culture medium.

Applications

Vegetative propagation

Tissue culture techniques are becoming increasingly popular as an alternative means of plant vegetative propagation, mass production of chemicals, and genetic engineering. The primary goal of plant tissue culture is crop management. This involves asexual methods of propagation to generate whole plants from single cells. In order for successful selection and genetic manipulation to occur, one must have successful *in vitro* plant regeneration, making this topic very important.

In vitro clonal propagation is referred to as micro propagation. In this context clonal means the production of genetically identical plants grown from parts of plants and reproduced by asexual reproduction. This procedure allows for many thousands of plants to be derived from a single cell or tissue in a relatively short amount of time. This appears to be advantageous over conventional plant reproduction in that only a small amount of plant tissue is required as the initial explant for regeneration of millions of clonal plants in one year. Therefore, this *in vitro* technique allows for speedy international exchange of plant materials. It also eliminates the danger of introducing disease to the crop if performed under sterile conditions. Furthermore, when crops are grown *in vitro* it allows for year round growth regardless of what Mother Nature throws at us. Unfortunately, it is very expensive to carry out clonal propagation due to the costly equipment and trained engineers to perform the procedures. In addition to high growth rates there could also be high contamination rates. Once contamination occurs it could spread through the crops causing great losses over a very short amount of time.

Biosynthesis of secondary metabolites

Cell cultures are also useful for the secondary metabolites they produce. Some of these metabolites that are a valuable source include flavors, natural sweeteners, industrial feed stocks, perfumes and commercial insecticides. These products do not perform vital physiological functions like amino acids or nucleic acids, but they are produced to ward off potential predators, attract pollinators, or combat infectious diseases. Another useful metabolite produced by plants includes Shikokin, which is a chemical used as both a dye and a pharmaceutical reagent. Plant cell suspension culture is valuable for studying the biosynthesis of secondary metabolites. Although there are limitations of cell culture systems in producing secondary metabolites, they are favored over conventional cultivation methods. This is because of their ability to produce useful compounds under controlled conditions as well as their capability of using this technique to produce chemicals to meet market demands. In addition, specific cells of a plant can be multiplied to produce a higher yield of specific metabolites, which cannot be done through conventional methods of cultivation.

Troubleshooting

Poor growth

Problems associated with standard plant tissue culture protocols are contamination of the cultures which often arise as a result of non-sterile techniques during the tissue dissection or contamination of an ingredient added to the tissue culture medium. If the cultures are initiated with a very high cell density, the dead and dying cells can lower the tissue culture pH and results in poor growth.

Insect cell culture

Introduction

Insect cell cultures are now commonly used in insect physiology, developmental biology, pathology, and molecular biology. As the advancement of tools became very rapid many protocols and standard procedures have been developed and the following protocol is one among them.

Principle

Cell Lines Used

SF-9
SF-21
High-Five

Sf21 is a cell line developed from ovaries of the fall armyworm, *Spodoptera frugiperda,* a moth species, which is an agricultural pest on corn and other grass species. It was originally developed in the United States at the Henry A. Wallace Beltsville Agricultural Research Center. Sf9 is a substrain (clone) of these cells that was isolated from Sf21 by researchers at Texas A&M University (Vanghn, 1977 and 1994).

Protocol

Media

For SF-9 and SF-21 cells:

TC-100 or Grace's media + 10% FBS (Fetal bovine serum) + supplements. For powdered media, dissolve in about 800 ml double distilled water.

If yeastolate and lactalbumin are not present, add: 3.3 g/L yeastolate and 3.3 g/L lactalbumin

For TC-100 media also add 0.03 g/L sodium bicarbonate

Adjust the pH to 6.2 with 1 M KOH (about 20-30 ml).

25 μg/ml gentamicin

contd...

Filter sterilize through a 0.22 micron filter.
Add heat inactivated FBS (10%)
For High-Five cells:
Ex-Cell 401 (liquid)
25 μg/ml gentamicin

Growth Characteristics : The optimal temperature for insect cell growth is 27°C. If an incubator is not available, the cells will grow at room temperature but at a slower rate. CO_2 incubators are not required. Cells may be grown in T-flasks or shaker flasks.

T-flasks: Grow cells to 80-100% confluency and then split 1:10. Remove adherent cells from the plastic surface of the flask by pipeting media with a wide bore, bent tip pipette. Overgrown cultures will have floating cells. Floating cells in a T-flask are an indicator of an unhealthy culture. It takes about one week after splitting for the cells to reach confluency. Re-use T-flasks once. It is not a good idea to use aliquots of these cells directly in experiments because 30-50% of them are trypan blue positive after being dislodged from the plastic surface.

Shaker flasks: Cells may be grown on orbital shaking platforms (rotating at about 100 RPM) at 27°C in sterile Erlenmeyer flasks. The flasks should be no more than 2/5 full (Maintain 10-15 ml of cells in a 50 ml flask). SF-21 and High-Five cells double in about 24 hours. An advantage of maintaining cells in shaker flasks is that you do not need to blow the cells off T-flasks which kill many of them. Use seed shaker flasks at 500,000 and 400,000 cells/ml for SF21 and High-Five, respectively, but more recently seed at 150,000 and 100,000 cells/ml, are used without problems. Add 10 U/ml heparin for the first two passages of High-Five cells to prevent clumping during adaptation to suspension culture from T flasks.

Split cells when they reach about 2×10^6 cells/ml. (High-Five cells suffer no adverse effects at 4×10^6cells/ml.) SF-21 cells should only be passaged about 10 times.

Cleaning Glass Shaker Flasks: Glass shaker flasks are bleached overnight, scrubbed with a brush and soap before cleaning in a dishwasher. Flasks are then soaked in distilled water overnight to remove residual chemicals. The water is then removed, and the flasks are autoclaved.

Freezing cells

Cells should be in exponential growth and more than 98% viable before freezing in liquid nitrogen.

1. Centrifuge the cells at 1,000 g for 5 min. Discard the supernatant and resuspend the cells in fresh media at a concentration of 1-2 X 10^7 cells/ml.
2. Mix equal volumes of resuspended cells and freezing medium [85% media, 15% DMSO-Di-(methyl sulfoxide)] and place on ice. Aliquot 1 ml volumes into cryovials.
3. Place cryovials into a Nalgene Cryo 1°C Freezing container (an isopropanol bath) and place directly at -70°C overnight.
4. Transfer cryovials into a liquid nitrogen storage tank.

Thawing Cells

1. Remove cryovials from liquid nitrogen and thaw rapidly (in about 1 minute) at 37°C. (Thawing High-Five cells slowly over about 30 minutes at room temperature is recommended)
2. Wipe the outside of the vial with 70% ethanol and transfer the cells to a T25 flask containing 4 ml complete media.
3. Wait 2-3 hours for the cells to adhere to the plastic. Remove the media and feed adherent cells with fresh media to dilute the DMSO. (High-Five cells adhere slowly to plastic, wait overnight before feeding the cells.) If cells are not adherent, they are probably dead. In this case, discard and thaw another vial of cells.
4. After 24 hours, feed the cells with fresh media.
5. For cell maintenance, follow the procedures in the Growth Characteristics/T-flask section.

Applications

Insect cell cultures are widely used in viral diagnosis and biotechnology, for the production of recombinant proteins, viral pesticides and vaccines as well as in basic research in genetics, molecular biology, biochemistry, endocrinology and virology.

Large number of cell lines from Dipteran, Hemipteran, and Lepidopteran insects were established and characterized in India. With the availability of the modern tools in molecular biology and the advancements made in biotechnology, the indigenous cell lines may prove useful in creating a future without biohazardous chemical pesticides as well as producing life saving pharmaceuticals and vaccines for many diseases.

Troubleshooting

Many of the problems that may arise are associated with standard tissue culture protocols. Contamination of the cultures is often a result of non-sterile techniques during the tissue dissection or contamination of an ingredient added to the tissue culture medium. If the cultures are initiated with a very high cell density, the dead and dying cells can lower the tissue culture pH and affects the culture.

Animal cell culture

Introduction

The use of animal and human cell cultures has become very beneficial for diverse applications in biotechnology and biomedical research. Originally used as substrates for the production of viral vaccines, animal and human cell cultures became an indispensable tool to produce a variety of products, including biopharmaceuticals, monoclonal antibodies and products for gene therapy. Although animal cell culture was first successfully undertaken by Ross Harrison in 1907, it was not until the late 1940's to early 1950's that several developments occurred that made cell culture widely available as a tool for scientists. First, there was the development of antibiotics that made it easier to avoid many of the contamination problems that plagued earlier cell culture attempts. Second was the development of the techniques, such as the use of trypsin to remove cells from culture vessels, necessary to obtain continuously growing cell lines (such as HeLa cells). Third, using these cell lines, scientists were able to develop standardized, chemically defined culture media that made it far easier to grow cells. These three areas combined to allow many more scientists to use cell, tissue and organ culture in their research.

Principle

Cells can be isolated from tissues for ex vivo culture in several ways. Cells can be easily purified from blood; however only the white cells are capable of growth in culture. Mononuclear cells can be released from soft tissues by enzymatic digestion with enzymes such as collagenase, trypsin, or pronase, which break down the extracellular matrix. Alternatively, pieces of tissue can be placed in growth media, and the cells that grow out are available for culture. This method is known as explant culture. Cells that are cultured directly from a subject are known as primary cells. With the exception of some derived from tumours, most primary cell cultures have limited lifespan. After a certain number of population doublings cells undergo the process of senescence and stop dividing, while generally retaining viability.

In the present chapter we describe the steps in detail to isolate and expand neural stem cells in the form of neurospheres from tissue dissections of the post-natal mouse brain (Pacey *et al.*, 2006). Procedures for the long term passage of neurospheres and the cryopreservation of neurospheres are also provided. In addition to the guidelines and tips for generating neurosphere cultures, we describe the method to prepare neurospheres for analysis by light microscopy. The ability to section neurospheres for histology or immunocytochemistry provides the researcher with the additional dimension to study specific molecular and cellular aspects of the neurosphere. Many of the commercial kits designed to assess aspects of the cell cycle, programmed cell death, and cell signaling can be readily applied to the outlined protocols.

Reagents

Reagent Setup

Distilled H_2O (dH_2O) sterilized with a 0.22 µm pore size filter.

See Table 1 for the recipes to make the tissue dissection solution and the Enzyme Mix.

Refer Table 2 for the components added to the Serum Free Medium (SFM).

The SFM is DMEM (Dulbecco's Modification of Eagle's medium) /F12 (1:1) + L glutamine and 15mM HEPES (N-(2-hydroxyethyl)-piperazine-N'-2-ethanesulfonic acid)

SFM is filtered with a 0.22 µm pore size filter after the addition of the components, with the exception of the growth factors (EGF, FGF), B-27 and ITSS which are added to the sterile SFM.

Additional components added to the SFM

Putrescine (100x stock) (1,4-Diaminobutane dihydro-chloride) - dissolve 0.096g in 100ml dH_2O and filter with a 0.22 µm pore size filter (store at 4°C).

Progesterone (1000x stock) - dissolve 0.00629g in 100 ml of dH_2O and filter with a 0.22 µm pore size filter (store at 4°C).

1.0M HEPES Buffer

B-27 Supplement

TIP: *In the past few years this supplement has often been on back order. Check in advance for availability.*

Insulin-Transferrin - Sodium Selenite Supplement (ITSS) - dissolved in 5.0 ml sterile dH_2O (1000x stock) and make 100µl aliquots (store at -20°C). One aliquot is added to the SFM.

Heparin - dissolve 0.05g in 2.0 ml of filtered dH_2O (store at 4°C)

Trypsin inhibitor - 1.0 mg/ml in SFM. Make fresh each time.

Growth factors

Fibroblast growth factor-basic (FGF) – 25 μg dissolved in 1.0 ml of 5mM Tris (pH 7.6) and sterilized with a 0.22μm pore size filter. Make 20 μl aliquots (stored at -20°C) and add 1 aliquot to 100 ml of SFM.

Epidermal growth factor (EGF)

0.1 mg dissolved in 1.0 ml of phosphate buffered saline containing 0.1% bovine serum albumin. Make 20 μl aliquots (stored at -20°C) and add 1 aliquot to 100 mls of SFM.

Critical

The biological activity of the growth factors in the SFM decreases with time. For optimal results, use SFM with the growth factors on the day of preparation. We keep our medium for up to 7 days for passage of the neurospheres.

Equipment

- CO_2 desiccation chamber
- Dissecting microscope
- Water bath (shaking) set at 37°C
- Micro-dissecting instruments
- Scapel blades (No. 10 curved)
- Betadine antiseptic solution (10% povidone-iodine)
- Cidex antiseptic (ortho-Phthaladehyde solution)
- Table top centrifuge
- 60 × 15 mm style polystyrene tissue culture dishes
- 24 well (multi-well) tissue culture plates
- 5 ml polypropylene round bottom tubes
- 15 ml polystyrene conical tubes
- 50 ml polypropylene conical tubes
- Tissue culture equipment
- Incubator at 37°C with 95% air and 5% CO_2
- Laminar air flow hood or biological safety cabinet

Time Taken

With all the reagents prepared in advance, the dissection and cell plating procedure will require 2-3 hours (depending on experience).

Protocol

Pre-surgery

1. Set the water bath to 37°C.
2. Bubble the tissue dissection solution with 95% O_2 and 5% CO_2 for 10 minutes.
3. Disinfect the dissection area with 95% ethanol.
4. Sterilize the dissection instruments in an antiseptic solution (Cidex followed by a complete sterile water rinse and then immerse in 95% ethanol).
5. Fill three 60 × 15 mm style tissue culture dishes with 3.0 ml of the tissue dissection solution.

TIP : *In addition to sterile micro-dissecting instruments, three sets of sterile instruments should be used. A separate set for the skin, skull and removal of the brain will reduce the chances of contamination as you proceed through the tissue layers. The sterile instruments can be dipped into sterile tissue dissection solution to avoid ethanol contact with the tissues.*

Removal and dissection of the brain

6. Euthanize the mouse in the CO_2 chamber. Animals less than 3 weeks old should initially be decapitated.
7. Shave the top of the head.
8. Swab the head with gauze soaked in Betadine to sterilize and remove any loose hairs from the skin.
9. Swab the head with gauze soaked in 95% ethanol.
10. Make a midline incision in the skin with a scalpel blade over the entire length of the skull.
11. Reveal the surface of the skull by reflecting the skin.

12. Make additional cuts in the skin to expose the sides of the skull just below the ears.
13. Cut the skull on both sides with small scissors (Remove the upper portion of the skull cap without cutting the brain).
14. Break off the skull with forceps.

    CAUTION : Make sure the dissecting instruments are free of ethanol before touching the brain. The ethanol will fix the tissue!

15. Slide curved forceps under the base of the brain to cut the spinal cord, blood vessels and cranial nerves that are connected to the base of the brain.

    TIP : *Apply slight downward pressure with the curved forceps on the floor of the skull under the brain. Slide the forceps back and forth to loosen the brain.*

16. Transfer the brain with the same curved forceps to a 60 mm tissue culture dish with the dissection solution.
17. Strip any remaining meninges from the surface of the brain, rinse thoroughly and place the brain into a second 60 mm tissue culture dish with dissection fluid.

    TIP : *The meninges are most readily stripped from the surface of the brain when viewed under the dissection microscope.*

18. Cut the brain along the midline and isolate the desired region of the brain with fine forceps.

    TIP : *While all regions of the brain will generate neurospheres, specific zones and regions will yield higher numbers of neurospheres. Dissections that concentrate on the cell layers that form the ventricles (mainly subventricular zone, the hippocampus and the rostral migratory stream) are areas that will yield the most neurospheres. A working knowledge of the mouse brain in combination with an atlas of the developing and postnatal mouse brain is an asset.*

19. Transfer the dissected tissue into a third 60 mm tissue culture dish containing tissue dissection solution.
20. Cut the tissue into small squares (approximately 1.0 mm) with No. 10 curved scalpel blades.

TIP : *The efficiency of cutting the tissue into small pieces is improved with experience and the aid of the dissecting microscope. Smaller pieces of tissue will dissociate better and yield a higher number of neurospheres.*

Dissociation of the brain tissue

21. Transfer the dish with the dissected tissue (small squares) to the tissue culture hood and use a Pasteur pipette to load the tissue pieces into a 5 ml polypropylene round bottom tube.
22. Centrifuge for 1 minute at 250g.
23. Remove the supernatant and add the enzyme mix (1.0 ml).

TIP : *If a specific brain region is not required, the entire brain from a mouse pup (less than 1 week old) can be used to generate the maximum number of neurospheres. In this situation, the entire brain is divided into 2 tubes, with each tube containing 1.0 ml of the enzyme mix. Younger animals will generate more neurospheres.*

24. Place the tube in the water bath for at least 40 minutes (newborn to one week old) or for approximately 1.5 hours (young adult less than 2 months old) and triturate the samples briefly with a Pasteur pipette every 20 minutes.

TIP : *Check the status of the tissue dissociation regularly. The exact amount of time required depends on the age of the animal. Animals several months old may require 2-3 hours of digestion. Cell viability is correspondingly reduced with prolonged digestion.*

CAUTION : *Newborn brain tissue is digested rapidly by the enzyme mix. Excessive digestion will significantly reduce cell viability.*

25. Centrifuge for 5 minutes at 250g and remove the supernatant.
26. Add 4.0 ml of the trypsin inhibitor.
27. Triturate (10 times) with a Pasteur pipette to break up the pellet and take care to minimize the introduction of air bubbles.
28. Place the tube in the water bath for 10 minutes at 37^0C.
29. Centrifuge for 5 minutes at 500g and remove the supernatant.
30. Re-suspend the cells in 0.5 ml of SFM and triturate sufficiently to produce a single cell suspension.

TIP : *A small bore fire polished Pasteur pipette is recommended for dissociating the suspension into single cells.*

Plating the cells

31. Calculate and adjust the viable cell concentration with a hemacytometer. Viable cells are determined by Trypan Blue dye exclusion.

TIP : *A plating density of 10-15 cells/μl will favor the establishment of neurospheres from single cells. If you do not require a clonal assay and desire a higher production of neurospheres, plate the cells at a higher density (eg. 50-100 cells/μl). Plating the cells at a higher density will also generate the neurospheres more rapidly with a shorter interval between the initial passages.*

32. Add 0.5 ml of SFM containing the cells to each well of a 24 multi-well plate.
33. Incubate at 37°C with 95% air and 5% CO_2.

Passage of neurospheres

TIP : *Check each well for contamination before pooling the contents of the entire plate, especially for the first passage. A heavily contaminated well will lower the pH with an obvious change in color (yellow) of the culture medium. It is not unusual to have a well (s) slightly contaminated. A low level of contamination will not be detected by a color change in the medium, but only detected by observation with a phase contrast microscope. If you miss a contaminated well, all the wells of the next passage will be contaminated!*

1. Transfer the neurospheres and medium from all wells to a 15 ml conical tube.
2. Centrifuge for 5 minutes at 200g.
3. Remove the supernatant and add 2.0 ml of TrypLE (Cell dissociation enzyme) to the tube.
4. Use a Pasteur pipette to mix the neurospheres with the TrypLE.
5. Place the tube in the water bath for 20 minutes at 37°C.
6. Centrifuge for 5 minutes at 500g.
7. Remove the supernatant and re-suspend the cells in 0.5 ml of SFM.
8. Triturate with a Pasteur pipette (60-70 times).

   TIP : *Avoid the formation of air bubbles. Excessive air bubbles will lower the cell viability and enhance the possibility of contamination.*

9. Determine the cell concentration by Trypan Blue dye exclusion and plate the desired concentration of cells to a new 24 multi-well plate. Incubate at 37°C with 95% air and 5% CO_2.

   TIP : *The best time to pass the neurospheres is often determined on an individual basis dictated by the experimental requirements and timetable. Passage of the neurospheres renews the SFM and expands the number of neurospheres. Plates containing an average of at least 25 neurospheres per well can be passed into 2 plates. With densities in excess of 50 neurospheres per well, 3 plates can easily be established from the initial 24 multi-well plate. A stage micrometer can be used to calibrate the crossed (focusing) lines on the ocular(s) of the phase contrast microscope and allow you to quickly determine the size of the neurospheres.*

Cryopreservation and re-establishment of neurospheres

1. Pool the culture medium with the neurospheres from a 24 multi-well plate into a 15 ml conical tube.
2. Centrifuge for 5 minutes at 200g.
3. Remove the supernatant and re-suspend the cells in 10.0 ml SFM (without EGF and FGF) containing 15.0% dimethyl-sulfoxide (DMSO).
4. Gently mix the neurospheres in the SFM with a Pasteur pipette. Aliquot 1.5 ml of the mixture into polypropylene cryovials.
5. Transfer the tubes to a -80°C freezer overnight.
6. The following morning, transfer the vials to a liquid nitrogen cryofreezer.

   TIP : *If you do not have access to a liquid nitrogen freezer, the neurospheres can be kept for several weeks at -80°C. The number of viable neurospheres obtained in culture will decrease with prolonged storage at -80°C.*
7. To re-establish the neurospheres in tissue culture, remove the cryovial from the freezer and let the vial warm up to room temperature on the bench.
8. Add the contents of the cryovial slowly to a 15 ml conical tube containing 10 ml of SFM.
9. Centrifuge for 5 minutes at 200g and remove the supernatant.
10. Gently mix the neurospheres with 4.0 ml of the SFM with a Pasteur pipette.
11. Fill 8 wells of a 24 multi-well plate with 0.5 ml of the mixture and incubate at 37°C with 95% air and 5% CO_2.

    TIP : *Smaller neurospheres will survive the freezing process better than larger neurospheres (in excess of 100μm in* *diameter). The viability of neurospheres following the freezing process in this tissue culture medium is low (< 20%). It is recommended that the individual wells contain at least 50-100 neurospheres. You want to freeze a high concentration of neurospheres to enhance neurosphere re-establishment after freezing. Neurosphere viability can also be enhanced by the addition of 20% fetal bovine serum (FBS) to the freezing medium. However, keep in mind that FBS will induce differentiation.*

Preparation of neurospheres for cryostat sectioning and immuno-cytochemistry

Procedures have been developed to section intact neurospheres.

1. Transfer the culture medium with neurospheres to a 50 ml polypropylene conical tube.
2. Allow the neurospheres to settle by gravity (approximately 30 minutes. The time for the neurospheres to settle is dependent on the size of the neurospheres.

CAUTION : *Centrifugation of the viable neurospheres will alter the shape of the neurospheres.*

3. Remove the supernatant, add and mix gently with 2.0 ml of 4.0% phosphate buffered paraformaldehyde (fixative).
4. Allow the neurospheres to settle in the tube.
5. Remove the fixative supernatant with a Pasteur pipette and rinse with 5.0 ml of phosphate buffered saline (PBS).
6. Allow the neurospheres to settle for at least 30 minutes before removal of the PBS.
7. Repeat the rinsing procedure at least 3 times. Immunocyto-chemistry may be performed on intact neurospheres at this stage.

TIP : *Observing the removal of the supernatant at any step under the dissecting scope ensures maximum withdrawal of the fluid without disturbing the cell pellet.*

8. Remove the final PBS rinse and add 5.0 ml of 30% sucrose (dissolved in 0.1M PBS).
9. Transfer the tube to the refrigerator and allow the neurospheres to settle overnight.
10. Remove the 30% sucrose and add embedding medium for frozen tissue specimens for 1 hour (Sakura Tissue-Tek O.C.T. compound)
11. Prepare the cryostat chuck with a layer of O.C.T.
12. Transfer the neurosphere pellet to the O.C.T. layer on the cryostat chuck.

CAUTION: *You can lose the neurosphere pellet quickly during the transfer of the pellet from the tube to the cryostat chuck. With a Pasteur pipette or an 1000 μl Eppendorf pipettor (using the blue tips) pre-load a small amount of O.C.T. into the tip and then withdraw the neurosphere pellet into the tip, continuous with the pre-loaded O.C.T. Try not to induce any air bubbles into the O.C.T.*

13. Cut sections at desired thickness and collect on APTEX (3-Aminopropyl triethoxy-silane) coated glass microscope slides.
14. Continue with the desired immunocytochemical procedure (non-specific blocking, application of primary antibody etc.).

Troubleshooting

Neurosphere generation

Many of the problems that may arise are associated with standard tissue culture protocols. Contamination of the cultures is often a result of non-sterile techniques during the tissue dissection or contamination of an ingredient added to the tissue culture medium. If the cultures are initiated with a very high cell density, the dead and dying cells can lower the tissue culture pH and prevent the progression of neurosphere development.

Cryopreservation and re-establishment of neurospheres

If low numbers of neurospheres are obtained after the freezing process, cut back on the number of wells seeded (4 or even 2 wells for the initial seeding step). Alternatively, you may wish to dissociate the neurospheres into single cells for freezing.

Critical Steps

- Step 4 - Removal and dissection of the brain
- Step 17 - The meninges and associated blood vessels are a source of contaminating cells (blood cells, endothelial cells and fibroblasts)
- Step 20 - Dissociation of the brain tissue
- Step 27 - Cryopreservation and re-establishment of neurospheres
- Step 8 - Add the contents of the cryovial *slowly* to the conical tube with the SFM
- Step 9 - Make sure all the freezing medium is removed from the pellet. The DMSO in tissue culture will kill the cells.

Table 1: Tissue Dissection Solutions (250 ml)

2.0 M NaCl	15.5 ml
1.0 M KCl	1.25 ml
1.0 M $MgCl_2$	0.80 ml
155 mM $NaHCO_3$	41.90 ml
1.0 M Glucose	2.50 ml
108 mM $CaCl_2$	0.23 ml
Distilled H_2O	188.0 ml
Filter and store at 4°C and bubble with 95% O_2/5% CO_2 for ten minutes	
Enzyme mix	
Trypsin	0.04 g
Type 1-S Hyaluronidase	0.02 g
Kynurenic Acid	0.004 g
Dissolve enzyme components in 30 ml of the tissue dissection solution and filter with 0.22µm filter	
Make 1.0 ml aliquots and store at -20°C	

Table 2: Serum free media (SFM 100 ml)

DMEM/F12	94.3 ml
30% Glucose	2.0 ml
1M HEPES Buffer	0.5 ml
Progesterone (100x)	0.1 ml
Putrescine (100x)	1.0 ml
B 27 Growth Supplement	2.0 ml
EGF	20 μl
FGF	20 μl
ITSS	100 μl
Heparin	7.32 μl
8ml of 15 % $NaHCO_3$ (W/V) must be added to each bottle of DMEM/F12 (500 ml) store at 4° C	

Applications

Cell culture has become one of the major tools used in cell and molecular biology. Some of the important areas where cell culture is currently playing a major role are briefly described below:

Model Systems

Cell cultures provide a good model system for studying 1) basic cell biology and biochemistry, 2) the interactions between disease-causing agents and cells, 3) the effects of drugs on cells, 4) the process and triggers for aging, and 5) nutritional studies.

Toxicity Testing

Cultured cells are widely used alone or in conjunction with animal tests to study the effects of new drugs, cosmetics and chemicals on survival and growth in a wide variety of cell types. Especially important are liver- and kidney-derived cell cultures.

Cancer Research

Since both normal cells and cancer cells can be grown in culture, the basic differences between them can be closely studied. In addition, it is possible, by the use of chemicals, viruses and radiation, to convert normal cultured cells to cancer causing cells. Thus, the mechanisms that cause the change can be studied. Cultured cancer cells also serve as a test system to determine suitable drugs and methods for selectively destroying types of cancer.

Virology

One of the earliest and major uses of cell culture is the replication of viruses in cell cultures (in place of animals) for use in vaccine production. Cell cultures are also widely used in the clinical detection and isolation of viruses, as well as basic research into how they grow and infect organisms.

Cell-Based Manufacturing

While cultured cells can be used to produce many important products, three areas are generating the most interest. The first is the large-scale production of viruses for use in vaccine production. These include vaccines for polio, rabies, chicken pox, hepatitis B and measles. Second, is the large-scale production of cells that have been genetically engineered to produce proteins that have medicinal or commercial value. These include monoclonal antibodies, insulin, hormones, etc. Third, is the use of cells as replacement tissues and organs. Artificial skin for use in treating burns and ulcers is the first commercially available product. However, testing is underway on artificial organs such as pancreas, liver and kidney. A potential supply of replacement cells and tissues may come out of work currently being done with both embryonic and adult stem cells. These are cells that have the potential to differentiate into a variety of different cell types. It is hoped that learning how to control the development of these cells may offer new treatment approaches for a wide variety of medical conditions.

Genetic Counseling

Amniocentesis, a diagnostic technique that enables doctors to remove and culture fetal cells from pregnant women, has given doctors an important tool for the early diagnosis of fetal disorders. These cells can then be examined for abnormalities in their chromosomes and genes using karyotyping, chromosome painting and other molecular techniques.

Genetic Engineering

The ability to transfect or reprogram cultured cells with new genetic material (DNA and genes) has provided a major tool to molecular biologists wishing to study the cellular effects of the expression of theses genes (new proteins). These techniques can also be used to produce these new proteins in large quantity in cultured cells for further study. Insect cells are widely used as miniature cells factories to express substantial quantities of proteins that they manufacture after being infected with genetically engineered baculoviruses.

Gene Therapy

The ability to genetically engineer cells has also led to their use for gene therapy. Cells can be removed from a patient lacking a functional gene and the missing or damaged gene can then be replaced. The cells can be grown for a while in culture and then replaced into the patient. An alternative approach is to place the missing gene into a viral vector and then "infect" the patient with the virus in the hope that the missing gene will then be expressed in the patient's cells.

Drug Screening and Development

Cell-based assays have become increasingly important for the pharmaceutical industry, not just for cytotoxicity testing but also for high throughput screening of compounds that may have potential use as drugs. Originally, these cell culture tests were done in 96 well plates, but increasing use is now being made of 384 and 1536 well plates.

References

Bez, A., Corsini, E., Curti, D., Biggiogera, M., Colombo, A., Nicosia, R.F., Pagano, S.F. & Parati, E.A. (2003) Neurosphere and neurosphere-forming cells: morphological and ultrastructural characterization. *Brain Res.* 993, 18-29.

Chawla, H.S. (2000) Introduction to Plant Biotechnology. Science Publishers Inc. Enfield, NH, USA.

Franklin, K.B.J. & Paxinos, G. (Eds.) (1977) The Mouse Brain in Stereotaxic Coordinates, Academic Press, San Diego.

Freshney, R. I. (1994) Culture of Animal Cells, A Manual of Basic Technique, 3rd edition, Alan R. Liss, Inc., New York.

Goldman, S. (2005) Stem and progenitor cell-based therapy of the human central nervous system. *Nature Biotechnol.* 23, 862-870.

Gritti, A., Bonfanti, L., Doetsch, F., Caille, I., Alvarez-Buylla, A., Lim, D.A., Galli, R., Veerdugo, J.M., Herrera, D.G. & Alvarez-Buylla, A. (2002) Multipotent neural stem cells reside into the rostral extension and olfactory bulb of adult rodents. *J. Neurosci.* 22, 437-445.

Jacoby, W. B. & Pasten, I. H. (1979) Methods in Enzymology: *Cell Culture*, 58, eds. Academic Press, New York.

John Paul. (1975) Cell and Tissue Culture, 5th edition, Churchill Livingstone, Edinburgh.

Kempermann, G., Gast, D., Kronenberg, G., Yamaguchi, M. & Gage F.H. (2003) Early determination and long-term persistence of adult-generated new neurons in the hippocampus of mice. *Development*, 130, 391-399.

King, P.J. (1984) Induction and maintenance of cell suspension cultures. In: *Cell Culture and Somatic Cell Genetics of Plants.* (Vasil, I.K. ed) 1, pp 130-138. Academic Press, New York.

Lois, C. & Alvarez-Buylla, A. (1993) Proliferating sub ventricular zone cells in the adult mammalian forebrain can differentiate into neurons and glia. *Proc. Natl. Acad. Sci.* USA, 90, 2074-2077.

Mather, J. & Barnes, D. (1998) Animal Cell Culture Methods, 57, eds. *Methods in Cell Biology*, Academic Press, San Diego.

Maslov, A.Y., Barone, T.A., Plunkett, R.J. & Pruitt, S.C. (2004) Neural stem cell detection, characterization, and age-related changes in the subventricular zone of mice. *J. Neurosci.* 24, 1726-1733.

Morshead, C.M., Garcia, A.D., Sofroniew, M.V. & van der Kooy, D. (2003) The ablation of glial fibrillary acidic protein-positive cells from the adult central nervous system results in loss of forebrain neural stem cells but not retinal stem cells. *Eur. J. Neurosci.* 18, 76-84.

Pacey, L.K.K., Stead, S., Gleave, J.A., Tomczyk, K. & Doering L.C. (2006) Neural Stem Cell Culture: Neurosphere generation, microscopical analysis and cryopreservation. *Nature protocol online.*

Pais, M.S., Mavituna, F., Novais, J.M. (1988) *Plant Cell Biotechnology.* Springer-Verlag Berlin Heidelberg, Germany.

Reynolds, B.A. & Rietze R.L. (2005) Neural stem cells and neurospheres - re-evaluating the relationship. *Nature Methods.* 2, 333-336.

Reynolds, B.A. & Weiss, S. (1992) Generation of neurons and astrocytes from isolated cells of the adult mammalian central nervous system. *Science,* 255, 1707-1710.

Rothblat G. H. & V. J. (1972) Cristofalo Growth, Nutrition and Metabolism of Cells in Culture eds. Volumes 1-3 by Academic Press, New York.

Suslov, O.N., Kukekov, V.G., Ignatova, T.N. & Steindler, D.A. (2002) Neural stem cell heterogeneity demonstrated by molecular phenotyping of clonal neurospheres. *Proc. Natl. Acad. Sci.*USA, 99, 14506-14511.

Vamling K. et Glimelius K. (1990) Legumes and Oilseed Crops I *in: Biotechnology in Agriculture and Forestry* 10, Springer Verlag.

Vaughn, J. L. (1994) Lepidopteran cell culture. In "*Arthropod Cell Culture Systems,*" (K. Maramorosch and A. H. McIntosh, Eds.), pp,37-50. CRC Press, Boca Raton, FL.

Vaughn, J. L., Goodwin, R. H., Tompkins, G. J., & McCawley, P. (1977) "The establishment of two cell lines from the insect *Spodoptera frugiperda* (Lepidoptera: Noctuidae)". *In Vitro.* 13, 213-217.

Vescovi, A.L., Reynolds, B.A., Fraser, D.D. & Weiss, S. (1993) bFGF regulates the proliferative fate of unipotent (neuronal) and bipotent (neuronal/astroglial) EGF-generated CNS progenitor cells. *Neuron*. 11, 951-966.

Waymouth, C. (1970) Osmolality of Mammalian Blood and of Media for Culture of Mammalian Cells *In Vitro*, 6: 109-127.

❑❑❑

Chapter 4

Nucleic Acid Extraction

DNA Isolation Protocol

Introduction

Most of the all procedures involved in molecular biology experiments require the purified form of nucleic acid. It forms a fundamental requirements for genome characterization, gene mapping and for identification and isolation of gene for genetic engineering. The most important step involved in removal of proteins which is carried out in different ways. The following protocol is one of the most commonly used protocols in animal, plant and microbial samples.

Principle

The nucleic acid is the fundamental requirement for most genomic characterization, gene mapping and for the identification and isolation of specific genes for genetic engineering. The principal involved in this procedure is that of the cell wall digestion in order to release cellular components when the tissue is homogenized in liquid nitrogen or dry ice, upon which the SDS or CTAB disrupt the cell membrane & nucleic acid get separated. This DNA is protected by endogenous nucleases by the addition of EDTA, which is a chelating agent that binds Mg^{++} ions, which is considered a necessary cofactor for most nucleases. Chloroform - phenol is used to denature and separate the proteins from DNA. The use of this mixture reduces the amount of aqueous solution retained in the organic phase and denatured protein forms a layer at the interface between the aqueous and organic phase. Isoamyl alcohol is used to separate these two layers, the RNA is degraded by RNAase available in the DNA solution and proteinase -K is used to break down protein molecule in to free amino acids over long incubation period.

Chemicals and buffers required for the extraction of nucleic acids

1. 1M Tris
2. 5M NaCl
3. 0.5M EDTA
4. Saturated phenol pH 8.0
5. Phenol : Chloroform : Isoamyl alcohol(24:24:1)
6. Chloroform
7. 2Pk/Lysis buffer
8. 3M Sodium acetate/ potassium acetate pH 5.2
9. 0.5M EDTA pH 8.0
10. TE buffer pH 8.0
11. Proteinase K (Final conc. 100 mg/ml)
12. RNAase (Final conc.100 mg/ml)
13. 70% alcohol

Buffer preparation protocols

1M Tris - pH 8.0

To prepare one liter of 1M Tris, dissolve 121.1 gm Tris in 500 ml of distilled water by using magnetic stirrer. Adjust pH 8.0 by adding HCl, than make up to one liter and autoclave.

2 PK buffer/ lysis buffer

TO prepare one liter 2 Pk buffer, dissolve 24.2 gm Tris (Final conc. 200 mM), 9.316 gm EDTA (Final conc. 25 mM EDTA), 17.52

gm Sodium Chloride (Final conc. 300 mM NaCl) and 20 gm SDS (2% Sodium Lauryl Sulphate) in 500 ml distilled water by using magnetic stirrer. Adjust pH 7.5 by adding HCl than make up to one liter and autoclave.

Phenol: Chloroform

Mix equal volume of phenol and chloroform. Equilibrate the mixture by extracting several tines with 0.1M Tris-HCl. Store the equilibrated mixture under an equal volume of 0.01M Tris-HCl at 4°C in dark bottles.

3M Sodium Acetate -pH 5.2

To prepare 1000 ml 3M Sodium Acetate, dissolve 408.24 gm Sodium Acetate in 800 ml of distilled water by using magnetic stirrer. Adjust pH 5.2 by adding Acetic acid than make up to 1000 ml. Dispense into aliquots and sterilize by autoclaving and stored at room temperature.

3M Potassium Acetate -pH 7.5

To prepare 1000 ml of 3M Potassium acetate, dissolve 294.42 gm Potassium acetate in 800 ml of distilled water by using magnetic stirrer. Adjust the pH 7.5 by adding acetic acid and then make up to 1000 ml. Divide the solution into aliquots and stored at -20°C.

5M Sodium Chloride

To prepare 100 ml 5M NaCl, dissolve 29.22 g NaCl in 100 ml distilled water, autoclave and stored at room temperature.

0.5M EDTA pH - 8.0

To make a 500mM or 0.5M solution of EDTA, add 181.6 g of $Na_2EDTA.2H_2O$ (di-sodium salt) to about 800 ml of distilled water and while stirring with a magnetic stirrer, adjust the pH to 8.0 with pellets of NaCl. Make the solution to 1000 ml with distilled water, sterilize by autoclaving and stored at room temperature.

20 mg/ml Proteinase K

Add 200 mg of Proteinase K to a 15 ml polypropylene tube containing 9.5 ml of H_2O. To reduce denaturation, always add protein to an aqueous solution instead of adding an aqueous solution to a protein. Gently rock the capped tube until the Proteinase K completely dissolved. Do not mix by overtaxing (this causes foaming, which indicates protein denaturation). Adjust the final volume to 10 ml. Divide in to aliquots and stored at -20° C. in general this solution is not sterilized.

10 mg/ml RNAase A

To avoid contamination of RNAase A, ware gloves and do not allow RNAase A to come in contact with any laboratory surface or equipment used for RNA wok. Dissolve 10 mg of pancreatic RNAase A in 1 ml of 10mM sodium acetate (pH 5.0). Place in a boiling -water bath for 15 minutes to inactivate any contaminating DNAase. Adjust pH to 7.5 with 1 M Tris-HCl. When cool, aliquot and stored at -20°. in general, this solution is not sterilized.

Cetyl Trimethyl-Ammonium Bromide (CTAB) stock

10g CTAB in 100 ml double distilled water, heat for dissolving completely, autoclave and stored at room temperature.

CTAB extraction buffer

100 mM Tris-HCl, pH 8.0 (0.5 ml from 1M stock), 10 mM EDTA, pH 8.0 (0.1 ml from 0.5 M stock), 1.4 M NaCl (1.4 ml from 5 M stock), 1% CTAB (0.5 ml from 10% stock) and 0.2% mercaptoethanol (0.01 ml) make up the volume to 5 ml (mercaptoethanol should be added just prior to use).

TE

To prepare 100 ml TE, add 1 ml of 1M Tris and 200 ml 0.5 M EDTA in 98.800ml autoclaved double distilled water.

70% Alcohol

Mix the 70 ml ethanol in 30 ml distilled water

Saturated Phenol

- The solidified distilled phenol from cold is kept out to bring to room temperature.
- The phenol is than melted on a boiling water bath.
- Equal volume of 0.5M Tris HCl pH 8.0 is added and stirred on magnetic stirrer for about 4-6 hour.
- Keep the mixture for about 6 hour in the cold room for separation of two layers.
- The above separated layer is removed.
- Again add equal volume of 0.5M Tris and stirred on magnetic stirrer for about 4-6 hour and keep this in cold room for about 4-6 hour, than above layer is separated.
- Again 1M Tris is added and keep stirring for 4-6 hours, keep in cold room for about 4-6 hour and the above layer is removed.

Phenol Chloroform Isoamyl alcohol method for insects

Most commonly used protocols for DNA extraction is Phenol: Chloloroform: Isoamyl-alcohol (PCI) method. It is a two step process that removes all the proteins and RNA's from the extracts. It is carried out by extracting the aqueous solution by once with saturated phenol, Phenol: Chloroform: Isoamyl alcohol and finally with Chloroform. Prior to extracting with the above organic solvents most of the proteins can be removed by digesting/ incubating with proteolytic enzymes like Pronase/Proteinase-K to a final concentration of 100mg/ml (Sambrook *et al.*, 1989).

The standard procedure is to remove protein from Nucleic acid. The aqueous solution (Nucleic acid) will tend to partition in to the organic phase if the pH of the phenol is not saturated to 8.0. Further more, although, phenol denatures protects efficiently it does not inhibit RNAase activity completely. These problems can be circumvented using a mixture of Phenol: Chloroform: Isoamyl alcohol. Subsequent extraction with chloroform further removes traces of phenol and any contaminants like fatty acids.

Phenol Chloroform Isoamyl alcohol method for insect DNA isolation

Protocol

STEP-1

1. Homogenate the whole insect in liquid nitrogen using a pestle and mortar.
2. Add sufficient quantity of 2PK buffer (2-3 ml per gram insect tissue) and transfer the sample into an Oakridge/centrifuge tubes.
3. Add proteinase K to a final concentration of 100 mg/ml and incubate the sample at 37°C for 3-4 hour or overnight in water bath.
4. Add equal volume of saturated phenol, mix gently, and allow to remain in room temperature for 5-10 min.
5. Centrifuge at 8000 rpm for 10-15 min. at room temperature.
6. Collect the supernatant into a clean autoclaved Oakridge tube.
7. Add equal volume of Phenol: Chloroform: Isoamyl alcohol (24:24:1) and repeat the step 5 and 6.
8. Add equal volume of chloroform and repeat step 5.
9. Collect the supernatant in an autoclaved glass culture tubes, add 1/10th volume of 3M sodium acetate pH 5.2 and 2 volume cold double distilled alcohol.
10. Allow for precipitation, by gentle mixing.
11. Collect the pellet, wash with 70% alcohol.
12. Dry the pellet in room temperature, dissolve in TE pH 8.0 and store at 4°C.

STEP II

RNAase treatment

The above extracted sample in TE solution may contain contaminants like traces of proteins and RNA which has to be eliminated before doing further experiments like PCR/Blotting etc. where purity of DNA plays an important factor.

1. To the above DNA sample add RNAase (100 mg/ml).
2. Incubate the sample at 37°C for 1hour in a water bath.
3. Add equal volume of Phenol: Chloroform: Isoamyl alcohol (24:24:1) and repeat the step 5 and 6 of step I.
4. Centrifuge at 8000 rpm for 10-15 min at room temperature.
5. Collect the supernatant into a clean autoclaved Oakridge tube.
6. Add equal volume of chloroform and repeat step 4 & 5
7. Collect the supernatant in an autoclaved glass culture tubes, add 1/10th volume of 3M sodium acetate pH 5.2 and 2 volume cold double distilled alcohol.
8. Allow for precipitation, by gently mixing.
9. Collect the pellet, wash with 70% alcohol.
10. Dry the pellet in room temperature, dissolve in TE pH 8.0 and store at 4° C.

Alternative method for insect DNA isolation from insect

DNA extraction by following proteinase K and SDS lysis method.

1. Wash the insect pest material in distilled water twice.
2. Homogenize in 150 ml lysis/2Pk buffer.
3. Add 5 ml of Proteinase K (final concentration 100µl/ml).
4. Incubate at 65° C for 30 minutes.
5. Add 100 ml of 3M Potassium acetate.
6. Incubate on ice for 1 hour.
7. Centrifuge at 14,000 rpm.
8. Transfer the supernatant in to a new eppendorf tube.
9. Precipitate the DNA with 2 volumes of ethanol.

10. Centrifuge for 10 minutes at 14,000 rpm.
11. Remove the ethanol supernatant.
12. DNA pellet was carefully washed with 400 ml ice-cold (-20° C) 70% ethanol.
13. Remove 70% ethanol and the sample was air dried for twenty minutes.
14. Re-suspend the DNA pellet in 100 ml Tris (10 mM Tris/HCl pH 8.0) and store at –20° C.

DNA isolation from plant materials

Usually CTAB method of nucleic extraction has been employed for plant DNA extraction (Murray and Thompson, 1980). The general protocol for plant DNA extraction using above method has been presented here under.

1. Plant material from the field, usually younger leaves should be taken and wraped in aluminium foil and stored immediately at -80° C or used for DNA extraction (avoid browning of material).
2. Crush the material with liquid nitrogen in a mortar and pestle until a fine mixture is prepared.
3. To mixture, add equal volume of CTAB extraction buffer (equal volume of the tissue) and mix thoroughly.
4. Incubate the homogenate following 150µg/ml proteinase K at 50° C for over night.
5. After incubation add equal volume of extraction buffer (chloroform-isoamyl alcohol 24:1). Mix thoroughly and followed by centrifugation at 15, 000 rpm for 5 minutes.
6. Transfer the upper aqueous phase to a new micro tube and add 1/10 volume of 10% CTAB solution and mix thoroughly.

7. Repeat the step 5 and add two volumes of ice cold absolute alcohol to the solution with slow mixing and keep the mixture at -20° C for one hour.
8. Centrifuge for 10 minutes at 5000 rpm at 4°C and discard the supernatant.
9. Wash the pellet with 70% ethanol and centrifuge at 5000 rpm for 5 minutes and discard the supernatant.
10. Dry the pellet and dissolve in TE (volume depends on size of the pellets)
11. Treat 100µg/ml RNAase and incubate at 37° C for one hour.
12. Add equal volume of Phenol-chloroform-isoamyl alcohol mixture (25:24:1) followed by mixing and centrifugation at 15000 rpm for 10 minutes at 4° C.
13. Collect the supernatant and add equal volume of chloroform-isoamyl alcohol (24:1) and centrifuged as above.
14. To the supernatant add 1/10 volume of 3M Sodium Acetate followed by two volumes of ice cold double distilled alcohol. Mix slowly and keep on ice.
15. Discard the supernatant following centrifugation, collect and wash the pellet with 70% alcohol.
16. The pellet was dried and dissolved in TE and stored at -20° C unit further use.

DNA isolation from microbes (*E. coli*)
(Sambrook *et al.*, 1989)

1. Grow *E. coli* culture overnight in rich broth.
2. Transfer 2 ml to a 2-ml micro centrifuge tube and spin for 2 min.
3. Decant the supernatant.
4. Drain well onto a Kim wipe.
5. Resuspend the pellet in 467 µl TE buffer by repeated pipetting.

6. Add 30 μl of 10% SDS and 3 μl of 20 mg/ml proteinase K, mix, and incubate for 1 hr at 37° C.
7. Add an equal volume of phenol/chloroform and mix well but very gently to avoid shearing the DNA by inverting the tube until the phases are completely mixed.
8. Carefully transfer the DNA/phenol mixture into a new eppendorf tube and spin at 12,000 rpm for 10 min.
9. Transfer the upper aqueous phase to a new tube and add an equal volume of phenol/chloroform.
10. Again mix well and transfer to a new Eppendorf tube and spin for 10 min.
11. Transfer the upper aqueous phase to a new tube.
12. Add 1/10 volume of 3M sodium acetate and mix.
13. Add 0.6 volumes of isopropanol and mix gently until the DNA precipitates.
14. Spool DNA onto a glass rod (or Pasteur pipette with a heat-sealed end).
15. Wash DNA by dipping end of rod into 1 ml of 70% ethanol for 30 sec.
16. Resuspend DNA in at least 200 μl TE buffer. Complete resuspension may take several days. Store DNA at 4° C for short term and at -20 or -80° C long term usage.

Applications

DNA carries the genetic information of a cell and consists of thousands of genes. Each gene serves as a recipe on how to build a protein molecule. Proteins perform important tasks for the cell functions or serve as building blocks. The flow of information from the genes determines the protein composition and thereby the functions of the cell.

Most of the experiments in molecular biology are based on DNA and a good quality and quantity is very much necessary.

Troubleshooting

Low yield

a) Incomplete homogenization or lysis of samples,

b) Incomplete solubilization of the final DNA pellet.

260/280 ratio < 1.70

a) Phenol was not sufficiently removed from the DNA preparation,

b) Acidic water was used for the spectrophotometric measurement.

DNA degradation

a) Tissues were not immediately processed or frozen after removing from animal,

b) Samples were homogenized with a Polytron or other high speed homogenizer.

RNA contamination

a) Too large volume of aqueous phase remained with the interphase and organic phase,

b) DNA pellet was not sufficiently washed with 10% ethanol-0.1 M sodium citrate solution.

RNA Isolation protocol (Guanidine thiocyanate/Phenol method)

Introduction

RNA is a polymer that contains ribose rather than deoxyribose sugars. The normal base composition is made up of guanine, adenine, cytosine, and uracil. When a cell receives a signal saying that a certain protein is needed, the code for producing protein is made. The DNA double helix unwinds and one strand of the helix becomes a template for producing the protein coding template. This template is a single strand of opposite bases (from DNA) and is called RNA (Ribonucleic Acid). Bases that are floating in the cell join up with opposite bases with Uracil taking the place of Thymine. This template is called mRNA (messenger RNA), because it serves as a code messenger between DNA and protein. The process of creating a mRNA from DNA is called transcription.

Principle

Ribose sugared RNA is a very renlitive Laboratory product. High levels of ribonucleaes, poly reach wides makes RNA very difficult to intract. As we know that RNA has special structural qualities extraction should be standardized very well before starting any expentant.

Reagents used

(Modification of Akop Seksanyan's protocol, which in itself is a modification of Chomczynski and Sacchi's protocol) (Sambrook *et al.* 1989).

An RNAase-free environment is critical

- Always wear clean gloves (don't touch your face/hair or "dirty" surfaces with your gloves and assume they remain nuclease-free; you exude nucleases).
- Use filter tips and other RNAase-free consumables.
- Make sure the pipette barrel is also nuclease-free.
- Use DEPC (diethyl pyrocarbonate)-treated water at all times.

Reagents (ensure sufficient volumes of all reagents are prepared in advance):

- GNTC (NB: Add 8μl of β-mercaptoethanol to every 1ml of GNTC solution before use!)
- 2M Sodium acetate, pH 4.0
- DEPC-phenol (RNAase-free phenol equilibrated in DEPC-water)
- Chloroform : Isoamyl alcohol (49:1)

Considerations

- Use at least 10µl of GNTC solution for every milligram of tissue.
- Homogenize the sample in GNTC as soon as possible; do not freeze sample prior to homogenization as the nucleases may be active before the GNTC/β-mercaptoethanol solution has the chance to inhibit nuclease activity.
- It may be necessary to use a larger volume to homogenize if using the Polytron. (1ml works well in a small 5ml snap-cap tube).

Protocol

Day One

1. After homogenizing the tissue in the appropriate amount of GNTC/β-mercaptoethanol solution, add the following (in the order written) and vortex between each addition:
   a. 0.1V of 2M Sodium Acetate, pH 4
   b. 1V of DECP-phenol
   c. 0.3V of Chloroform : Isoamyl-alcohol (the solution should turn milky by the end)
2. Incubate the mix on ice for 20 minutes; two layers should form.
3. While waiting for the layers to separate, ensure the centrifuge is at 4°C by doing a "fast-cool" spin. Also prepare collection tubes for the next step.
4. Centrifuge at 14,000 rpm (20,000 x *g*) for 25-30 minutes at 4°C.
5. Carefully transfer the top (aqueous) layer into the new tubes. Be VERY careful at this step not to pipette over any of the interphase. If necessary, only take 70-80% of the estimated aqueous layer.
6. Add 1.1V of isopropanol (2-propanol) to each tube.

7. Incubate overnight at -20°C to precipitate. (If absolutely necessary, precipitate at -80°C for 2 hours. Do not allow the sample to freeze).

Day Two

8. Again, ensure the centrifugation at 4°C by doing a "fast-spin".
9. Centrifuge the samples at 14,000 rpm for 25-30 minutes.
10. Discard the supernatant very carefully. Depending on the tube, the pellet may be very loose. Work quickly by tipping most of the solution out, and follow-up with a quick spin, then remove the remnant with a fine gel-loading tip.
11. Add 70% ethanol (use the same volume as of the isopropanol from the previous step).
12. Centrifuge at 14,000 rpm for 25-30 minutes at 4°C.
13. As before, work quickly to remove the supernatant. Follow up with a second spin and carefully remove the remnant with a fine tip.
14. Add 35μl of nuclease-free water.

NB: *If sufficient care was taken during pipetting of the aqueous phase, and the interphase was completely avoided, there should be no DNA in the sample. However, when extracting RNA from very small samples, the interphase is not always clearly visible, and thus carry-over of DNA often occurs. Should that happen, treatment with VERY clean DNAse can cure all known evils (OK, it can't, but it helps). We use Turbo DNase from Ambion, and have not yet had any problems. Of course, one should always check to see if one has DNA contamination in the first place, for example by doing a regular PCR (no RT!) using the RNA as a template, or by running it on a gel/RNA analysis chip.*

Optional DNAase treatment

15. To treat with DNAase I, add 4µl of DNAase buffer and 1µl of DNAase I
16. Incubate at 37° C for 45 minutes to one hour
17. At this point, either use the DNAase-inhibitor resin included in the kit[1] OR repeat the extraction by adding (don't forget to vortex between each addition):
    a. 500µl GNTC/β-mercaptoethanol
    b. 50µl Sodium Acetate
    c. 500µl DEPC-phenol
    d. 150µl Chloroform : Isoamyl-alcohol
18. Incubate for 20 minutes on ice to allow layers to separate. Check the centrifuge at 4°C and prepare tubes for the next step
19. Centrifuge at 14,000 rpm for 25-30 minutes at 4°C
20. Transfer the top (aqueous) layer into fresh tubes
21. Add 1.1V isopropanol
22. Incubate at -20°C overnight to precipitate

Day Three

23. Again, ensure the centrifuge is at 4°C by doing a "fast-spin"
24. Centrifuge the samples at 14,000 rpm for 25-30 minutes
25. Discard the supernatant very carefully. Depending on the tube, the pellet may be very loose. Work quickly by tipping most of the solution out, and follow-up with a quick spin, then remove the remnant with a fine gel-loading tip
26. Add 70% ethanol (use the same volume as of the isopropanol from the previous step)
27. Centrifuge at 14,000 rpm for 25-30 minutes at 4°C
28. As before, work quickly to remove the supernatant. Follow up with a second spin and carefully remove the remnant with a fine tip

29. Add nuclease-free water
    a. GNTC recipe (weights and volumes for 150ml given below)
    b. 4M Guanidine thiocyanate (70.9g / 150ml)
    c. 25mM Sodium citrate (1.1g / 150ml)
    d. 0.5% Sodium lauryl sarcosinate (2.5ml/g/150ml)
    e. -pH to 7.0, filter-sterilize and store at 4° C

    **And don't forget to add 8ml of β-mercaptoethanol for every 1ml of GNTC just before using it. (β-mercaptoethanol is light-sensitive).*

Applications

RNA is essential for protein synthesis, or the making of proteins in living creatures. Without RNA, humans and other living creatures could never exist.

Recently, RNA interference (RNAi) is a powerful tool to down-regulate single genes or sets of genes in a sequence-specific manner. Since its discovery in the 1990s, an impressive amount of research has revealed many of the key steps and players that produce the small RNA species at the centre of the mechanism of RNAi (i.e., small interfering RNAs and microRNAs).

RNA silencing- The first transgenic crop to be released for commercial growth made use of the antisense technology, now known to be based on RNA silencing. This was Calgene's Flavr Svr™ tomato, which was approved by the United States Food and Drug Administration in 1994 but was discontinued soon afterwards due to marketing problems and customer rejection.

Troubleshooting

Low yield

a. Incomplete homogenization or lysis of samples,
b. Incomplete solubilization of the final RNA pellet.

260/280 ratio < 1.6

a. Too small volume of the reagent used for sample homogenization,

b. Acidic water was used for the spectrophotometric measurement,

c. Contamination of the aqueous phase with phenol phase,

d. Incomplete solubilization of the final RNA pellet.

RNA degradation

a. Tissues were not immediately processed or frozen after removing from animal,

b. Samples used for isolation or the isolated RNA preparations were stored at -20° C instead of at -70° C,

c. Cells were dispersed by trypsin digestion,

d. Aqueous solutions or tubes used for solubilization of RNA were not RNAase-free,

e. Formaldehyde used for agarose-gel electrophoresis had a pH below 3.5.

DNA contamination

a. Too small volume of the reagent was used for sample homogenization,

b. Samples used for the isolation contained organic solvents (ethanol), strong buffers or alkaline solution,

c. Phase separation was performed at temp above 10° C.

References

Chomczynski, P.A. (1993) Reagent for Single-Step Simultaneous Isolation of RNA, DNA and Proteins from Cell and Tissue Samples. *BioTechniques.* 15, 532–536.

Maniatis T, Fritsch EF, Sambrook J (1982) *Molecular cloning: A laboratory manual.* Cold Spring Harbor Laboratory, Cold Spring Harbor, NY.

Murray HC, Thompson WF (1980) Rapid isolation of high molecular weight DNA. *Nucl.. Acids Res* 8:4321-4325.

Sambrook J, Fritsch EF, Maniatis T (1989) *Molecular cloning: a laboratory manual. Second edition.* Cold Spring Harbor Laboratory Press, New York, USA.

Sambrook J & Russel, D.W. (2001). *Molecular Cloning: A Laboratory Manual 3rd Ed.* Cold Spring Harbor Laboratory Press, NY.

Turpen, T.H. & Griffith, O.M. (1986) Rapid isolation of RNA by guanidinium thiocyanate/cesium chloride gradient method. *BioTechniques.* 4, 11–15.

Vareli, K., & Frangou-Lazaridis, M. (1996) Modification of the acid guanidinium thiocyanate-phenol-chloroform method for nuclear RNA isolation. *BioTechniques.* 21, 236–237.

Wang, H.M., Qi. & Cutler, A. (1993) A simple method of preparing plant samples for PCR. *Nucl. Acid Res.* 21, 4153-4154.

Zheng, K., Subudhi, P.K., Domingo, J., Magpantay, G. & Huang, N. (1995) Rapid DNA isolation for marker assisted selection in rice breeding. *Rice Genetics Newsletter.* 12, 255–258.

❑❑❑

Chapter 5

Determination of Quality and Quantity of Nucleic Acids

Introduction

In order to use a fine quality of DNA for various molecular biological work like sequencing, cDNA synthesis and cloning, RNA transcription, transfection, nucleic acid labeling (random primer labeling), etc., quality and quantity of DNA is extremely important to have the defined template concentration. Failure to produce results from the above techniques can sometimes be attributed to an incorrect estimate of the DNA template used.

Principle

After isolation of DNA, quantification and analysis of quality are necessary to ascertain the approximate quantity of DNA obtained and the suitability of DNA sample for further analysis. This is important for many applications including digestion of DNA by restriction enzymes or PCR amplification of target DNA. The most commonly used methodologies for quantifying the amount of nucleic acid in a preparation are: *(i)* Gel electrophoresis and *(ii)* Spectrophotometer analysis. If the sample amount is less, the former method is usually preferred.

Agarose gel electrophoresis for DNA quantification and quality Analysis

This method of quantification is based on the ethidium bromide fluorescent staining of DNA. Ethidium bromide is a fluorescent dye, which intercalates between the stacked bases. The fluorescent yield of the dye: DNA complex is much greater than the unbound dye. UV irradiation at 254nm is absorbed by the DNA and transmitted to the dye and the bound dye itself absorbs radiation at 302nm and 366nm. This energy is retransmitted at 590nm, the reddish-orange region of the visible spectrum. In case of plant genomic DNA; the nucleic acids are electrophoretically separated on a 0.7-0.8% agarose gel containing ethidium bromide at a final concentration of 0.5 mg/ml. The quantity of DNA can be estimated by comparing the fluorescent yield of the samples with a series of standards, for instance, lambda *(i)* DNA at varying known concentrations. This provides a very rapid and sensitive means of estimating the nucleic acid concentration. A large number of samples with as little as 1-5ng of DNA can be quantified. Besides quantification, it also allows provides the advantage of analyzing the quality of the DNA preparation. Native DNA, which migrates as a tight band of high molecular weight (340 kb), presence of RNA, and degraded/sheared DNA, if any, can be visually identified on the gel (Sambrook *et al.*, 1989).

Protocol

Prepare a 0.8% agarose gel.

Add 1 ml of 6X gel loading dye to 2-3 ml of each DNA sample before loading the wells of the gel. Addition of dye allows us to note the extent to which the samples might have migrated during electrophoresis, so that it can be halted at an appropriate stage.

Load at least 1 or 2 wells with uncut, good quality l DNA or any previously quantified DNA samples (50ng and 100ng) as molecular weight standards.

Run the submarine electrophoretic gel at 70V till the dye has migrated one-third of the distance in the gel.

DNA can be visualized using a UV transilluminator and quantified in comparison with the fluorescent yield of the standards.

Spectrophotometer method of determination

Analysis of UV absorption by the nucleotides provides a simple and accurate estimation of the concentration of nucleic acids in a sample. Purines and pyrmidines in nucleic acid show absorption maxima around 260nm (eg., dATP: 259nm; dCTP: 272nm; dTTP: 247nm) if the DNA sample is pure without significant contamination from proteins or organic solvents. The ratio of OD_{260}/OD_{280} should be determined to assess the purity of the sample. This method is however limited by the quantity of DNA and the purity of the preparation. Accurate analysis of the DNA preparation may be impeded by the presence of impurities in the sample or if the amount of DNA is too little. In the estimation of total genomic DNA, for example, the presence of RNA, sheared DNA etc. could interfere with the accurate estimation of total high molecular weight genomic DNA (Sambrook *et al.*, 1989).

Protocol

Take 1 ml TE buffer in a cuvette and calibrate the spectrophotometer at 260nm as well as 280nm.

Add 10 ml of each DNA sample to 900ml TE (Tris-EDTA buffer) and mix well.

Use TE buffer as a blank in the other cuvette of the spectrophotometer.

Note the OD_{260} and OD_{280} values on spectrophotometer.

Calculate the OD_{260}/OD_{280} ratio.

Comments

A ratio between 1.8-2.0 denotes that the absorption in the UV range is due to nucleic acids.

A ratio lower than 1.8 indicates the presence of proteins and/ or other UV absorbers.

A ratio higher than 2.0 indicates that the samples may be contaminated with chloroform or phenol. In either case (<1.8 or >2.0) it is advisable to re-precipitate the DNA.

The amount of DNA can be quantified using the formula:

$$\text{DNA concentration (mg/ml)} = \frac{OD_{260} \times 100 \text{ (dilution factor)} \times 50 \text{ mg/ml}}{1000}$$

Spectrophotometric conversions for Nucleic Acids

- A 260 of double stranded DNA = 50 mg/ml
- A 260 of single stranded oligonucleotides = 33 mg/ml
- A 260 of single stranded RNA = 40 mg/ml

Applications

Pure nucleic acid is necessary for the molecular biology experiments and with out it all experiments will fail or yield false results

Troubleshooting

Insufficient template DNA

Check preparation method and solutions.

Quantify DNA and add between 10-50ng to each PCR reaction.

Re-prepare DNA using recommended Alkali lysis method.

DNA poor quality or PCR inhibitor present

Check preparation method, solutions and storage.

Dilute extracted DNA 1 in 5 and re-test

Re-prepare DNA using recommended method and retest.

Excess template DNA

Quantitate and/or dilute DNA or re-prepare DNA using recommended method.

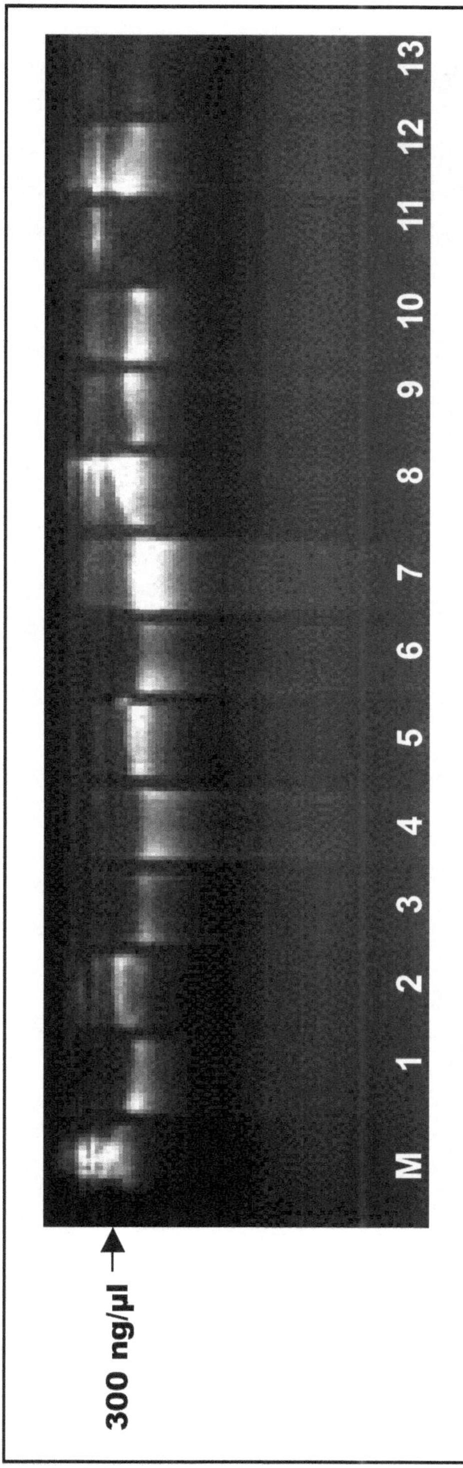

Figure 1: Gel picture of unamplified genomic DNA samples with standard marker. M- Marker (300 ng/µl), 1-13- DNA samples.

References

Hoisington, D. Khairallah, M. & Gonzalez-de-Leon, D. (1994). Laboratory Protocols: CIMMYT Applied Biotechnology Center. Second Edition, Mexico, D.F.: CIMMYT.

Sambrook J, Fritsch EF, Maniatis T (1989) *Molecular cloning: a laboratory manual. Second edition.* Cold Spring Harbor Laboratory Press, New York, USA.

❑❑❑

Chapter 6

Polymerase Chain Reaction

Introduction

This chapter gives first hand information on PCR for beginners. PCR is used to amplify a short, well-defined part of a DNA strand. This can be a single gene, or just a part of a gene. As opposed to living organisms, the PCR process can copy only short DNA fragments; usually up to 10 kb (*kb* stands for kilo base pairs). Certain methods can copy fragments up to 40 kb in size, which is still much less than the chromosomal DNA of a *eukaryotic cell*-for example, a human cell contains about three billion base pairs (Kary Mullis, 1983).

Principle

The PCR reaction is carried out in a thermal cycler. This is a machine that heats and cools the reaction tubes within it to the precise temperature required for each step of the reaction. Required DNA which is to be amplified will be loaded to the PCR well and ingredients like primer, dNTPs, PCR buffer, Taq enzyme and water are added. To prevent evaporation of the reaction mixture (typically volumes between 15-100μl per tube), a heated lid is placed on top of the reaction tubes or a layer of oil is put on the surface of the reaction mixture.

Reagents used

PCR requires several basic components. These components are:

- DNA template, or cDNA, which contains the region of the DNA fragment to be amplified
- Two primers, which determine the beginning and end of the region to be amplified
- Taq polymerase, which copies the region to be amplified
- Nucleotides, from which the DNA-Polymerase builds the new DNA
- Buffer, which provides a suitable chemical environment for the DNA-Polymerase

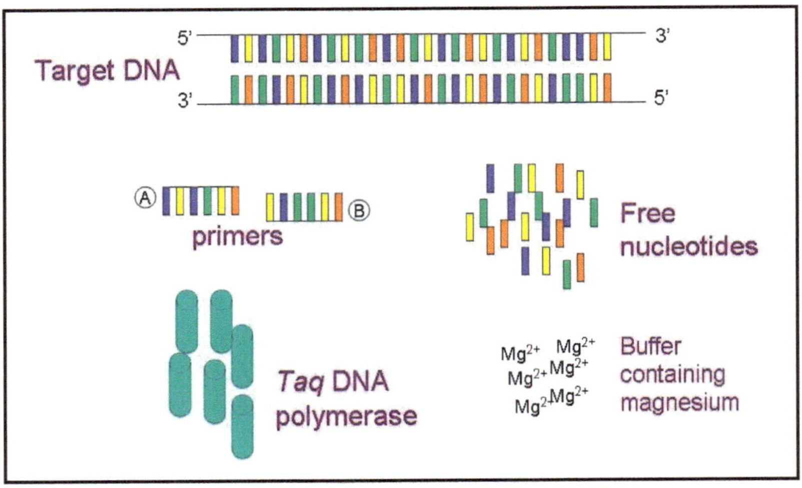

Figure 2: Basic components of PCR

Template DNA

The DNA to be amplified is extracted from the required tissue from plant or animal and it is to be quantified for needed concentration for the PCR amplification.

Primer

The DNA fragment to be amplified is determined by selecting primers. Primers are short, artificial DNA strands – not more than fifty (usually 18-25 bp) nucleotides that are complementary to the beginning and end of the DNA fragment to be amplified. They *anneal* (adhere) to the DNA template at these starting and ending points, where the DNA-Polymerase binds and begins the synthesis of new DNA strand at specified temperature.

dNTPs

Free deoxynucleotides (dNTPs) are added to the reaction to provide the building blocks of the new DNA being synthesized.

PCR Buffer

Reaction buffer is also added which provides the optimal conditions under which Taq polymerase works.

Presence of magnesium ions helps in catalytic action of enzymes, especially enzymes utilizing ATP, or those which use other nucleotides to synthesize DNA and RNA. 3

Taq DNA polymerase

Taq is isolated from *Thermus aquaticus,* a bacterium that grows in hot springs (75° C). Most enzymes would be killed at 95° C but this enzyme can with stand high temperature, so they can survive cycling through the high temperatures.

Protocol

You will make a master mix in an eppendorf tube, which then will be aliquot into separate PCR tubes.

The Component	Amount/reaction (20µl)	What it does?
10X PCR Buffer	2 µl	Controls pH for enzyme activity and primer binding
$MgCl_2$	2 µl	Enzyme activity, complexes dNTPs
dNTP	2 µl	Helps in the synthesis of new strands
Forward Primer	0.5 µl	Short oligonucleotide that binds to template. Goes in 5′ to 3′direction
Reverse Primer	0.5 µl	Short oligonucleotide that binds to template. Goes in 5′ to 3′ direction but on other strand reverse direction.
1% BSA	2 µl	Reduces nonspecific binding
Taq DNA polymerase	0.1 µl	Enzyme that copies DNA
DEPC Water	9.9 µl	Increases amount of reaction and decreases concentration of ingredients.

Once the master mix is made, you will pipette 19μl into a tube. Then you will add 1μl of DNA. This tube will then be placed in the thermocycler.

Thermocycler parameters: (Depends on the type of a amplification *i.e.* species used)

Initial Denaturation	95°C	10 minutes (depends on sample)
Denaturation	95°C	1 minute (depends on sample)
Annealing	55°C	1 minute (depends on sample)
Extension	72°C	1 minute (depends on sample)
Repeat	Denaturation, Ann. Ext	25 cycles (depends on sample)
Final Extension	72°C	10 minutes (depends on sample)
Hold	4°C	Forever

Adjusting of DNA concentration in samples for PCR

Normalization of DNA concentration of all the samples to a relatively equal level by appropriate dilutions when necessary. Based on approximate concentration assessed, prepare a table as follows:

- Make the dilution by first taking the appropriate amount of ddW or TE in fresh Eppendorf tubes and then adding the required volume of respective DNA from the stock solutions
- Mix well by tapping and quick spinning
- Run the required % agarose gel, check as in the section 4.1, load 5 μl of the diluted DNA solution (working solution) per sample
- Assess the gel image for relative intensity of genomic DNA band in each sample

Parameters	Sample
Assessed concentration of DNA stock Solution (ng of DNA/ µl)	100
Final concentration preferred in working solution (ng of DNA/ µl)	10
Folds of dilution needed (e.g.: 1:1, 1:2, 1:5, etc.)	1:10
Amount of ddW or TE to be used for dilution (µl)	90
Amount of DNA stock solution to be taken (µl)	10
Total volume of working solution with desired DNA concentration (µl)	1000

Applications

Application of PCR techniques in Molecular Ecology

The preservation of genetic diversity both within and among natural populations is a fundamental goal of conservation biology. Genetic diversity has been shown to identify populations that may benefit from the addition of relocated individuals or that should remain as they are because they represent different subspecies. Various molecular markers and PCR techniques have been used in a wide variety of ways, in a large number of studies, in an effort to attain this diversity.

RT PCR (Reverse transcriptase PCR)

The new forms of PCR respond to scientific and technical needs. Thus RT (Reverse Transcriptase) PCR permits scientists to amplify RNA in the same way that they copy DNA. Specialized enzymes allow research teams to amplify strands of DNA several thousand bases in length. Others enable copying with very high fidelity. A form of the technology that works in real time extends scientists' ability to undertake truly quantitative measurements. Technical advances include "hot start" systems and automatic cycling systems that make for more precise copying of DNA.

Amplify DNA for Cloning

PCR had an immediate impact on cloning technology. It could produce large quantities of DNA that could be readily cloned and subsequently used to study the functions and behavior of genes in living systems.

DNA sequencing reaction

The Sanger method is more commonly used for sequencing because it has been proven technically easier to apply, and, with the advent of PCR and automation of the technique, is easily applied to long strands of DNA including some entire genes. This technique is based on chain termination by dideoxy nucleotides during PCR elongation reactions. In this method, the DNA strand to be analyzed is used as a template and DNA polymerase is used, in a PCR reaction, to generate complimentary strands using primers. Four different PCR reaction mixtures are prepared, each containing a certain percentage of dideoxynucleoside triphosphate (ddNTP) analogs to one of the four nucleotides (ATP, CTP, GTP or TTP). Synthesis of the new DNA strand continues until one of these analogs is incorporated, at which time the strand is prematurely truncated.

Mapping genes and Linkage analysis

Gene mapping" refers to the mapping of genes to specific locations on chromosomes. It is a critical step in the understanding of genetic diseases. The genetic mapping is based on the linkage between "loci" (locations of genes). If two loci are usually inherited together, they are said to be "linked". Two loci on different chromosomes are not linked, because they are usually separated by independent assortment. A locus (singular of loci) may have different sequences, referred to as alleles. PCR plays a crucial role in detecting these sequence variation between genomes.

Population genetics

PCR technology promises advances in human genetics. This aims to discover the inherited mutations that occur in miniscule proportions in the population. Population genetists believe that this is essential to discovering disease-causing mutations in human populations.

Forensic science

Fields of science beyond traditional molecular biology laboratories have benefited from PCR. It has become a well-recognized tool in forensic science. Police labs routinely use it to identify blood and other forms of evidence

Archaeology

PCR has also become a useful tool in some unexpected scientific disciplines. For example, archaeologists have found it effective to determine relationships between ancient civilizations and to study the evolutionary biology of different animal species. PCR can amplify very small samples of DNA from virtually any tissue, including examples thousands of years old. These molecular readings have become very important in validating (and sometimes disproving) scientific conclusions based on circumstantial evidence.

Rapid amplification of template DNA for screening of uncharacterized mutations

Because of its rapidity and simplicity, PCR is ideally suited to providing numerous DNA templates for mutation screening. Partial DNA sequences, at the genomic or the cDNA level, from a gene associated with disease, or some other interesting phenotype, immediately enable gene-specific PCR reactions to be designed. Amplification of the appropriate gene segment then enables rapid testing for the presence of associated mutations in large numbers of individuals. By contrast, cell-based DNA cloning of the gene from numerous different individuals is far too slow and labor-intensive to be considered as a serious alternative.

Rapid genotyping for polymorphic markers

Restriction site polymorphisms (RSPs) result in alleles possessing or lacking a specific restriction site. Such polymorphisms can be typed using Southern blot hybridization.

A DNA probe representing the locus is hybridized against genomic DNA samples that have been digested with the appropriate restriction enzyme and size-fractionated by agarose gel electrophoresis. The resulting RFLPs have two alleles corresponding to the presence or absence of the restriction site. As a convenient alternative to RFLPs, PCR can type RSPs by simply designing primers using sequences which flank the polymorphic restriction site, amplifying from genomic DNA, then cutting the PCR product with the appropriate restriction enzyme and separating the fragments by agarose gel electrophoresis.

Degenerate oligonucleotide-primed PCR

It is a form of PCR which is deliberately designed to permit possible amplification of several products. The two primers may be *partially degenerate* oligonucleotides, composed of panels of oligonucleotide sequences that have the same base at certain nucleotide positions, but are different at others. As a result, there may be comparatively many primer binding sites in the source DNA. This provides a means of searching for a new or uncharacterized DNA sequence that belongs to a family of related sequences either within or between species. *Note:* the use of such primers also provides a way of cloning a gene when only a limited portion of amino acid sequence is known for the product.

Linker-primed PCR (ligation adaptor PCR)

Another way of enabling amplification of essentially all DNA sequences in a complex DNA mixture involves first ligating a known sequence to all fragments. To do this, the target DNA population is digested with a suitable restriction endonuclease, and double-stranded oligonucleotide linkers (also called adaptors) with a suitable overhanging end are ligated to the ends of target DNA fragments. Amplification is then performed using oligonucleotide primers which are specific for the linker sequences. In this way, all fragments of the DNA source which are flanked by linker oligonucleotides can be amplified.

Anchored PCR

It is often desirable to be able to amplify previously uncharacterized DNA sequences that neighbor a known DNA sequence, either at the genomic or cDNA level. To do this a form of anchored PCR is used. One of the primers is specific for the target sequence and the second primer is specific for a common sequence that can be introduced in different ways, such as by using a linker-primer method as described in the previous section, or by using primers that are modified at the 5' end so as to introduce a novel sequence.

Troubleshooting

Longer unspecific products

Decrease annealing time, Increase annealing temperature, Decrease extension time, decrease extension temperature to 62-68° C, Increase KCl (buffer) concentration to 1.2x-2x, but keep $MgCl_2$ concentration at 1.5-2mM. Increase $MgCl_2$ concentration up to 3-4.5 mM but keep dNTP concentration constant. Take less primer, take less DNA template and take less Taq polymerase.

Shorter unspecific products

Increase annealing temperature, Increase annealing time, Increase extension time, Increase extension temperature to 74-78° C, Decrease KCl (buffer) concentration to 0.7-0.8x, but keep $MgCl_2$ concentration at 1.5-2mM, Increase $MgCl_2$ concentration up to 3-4.5 mM but keep dNTP concentration constant, Take less primer, Take less DNA template and take less Taq polymerase.

Reaction working before not now

Make sure all PCR ingredients are taken in the reaction (buffer, template, Taq, etc) Change the dNTP solution (very sensitive to cycles of thawing and freezing, especially in multiplex PCR). If you just bought new primers, check for their reliability (bad primer synthesis?). Increase primer amount or increase template amount. Decrease annealing temperature by 6-10° C and check if you get any product. If you don't, check all your PCR ingredients. If you do get products (including unspecific ones) reaction conditions as described above.

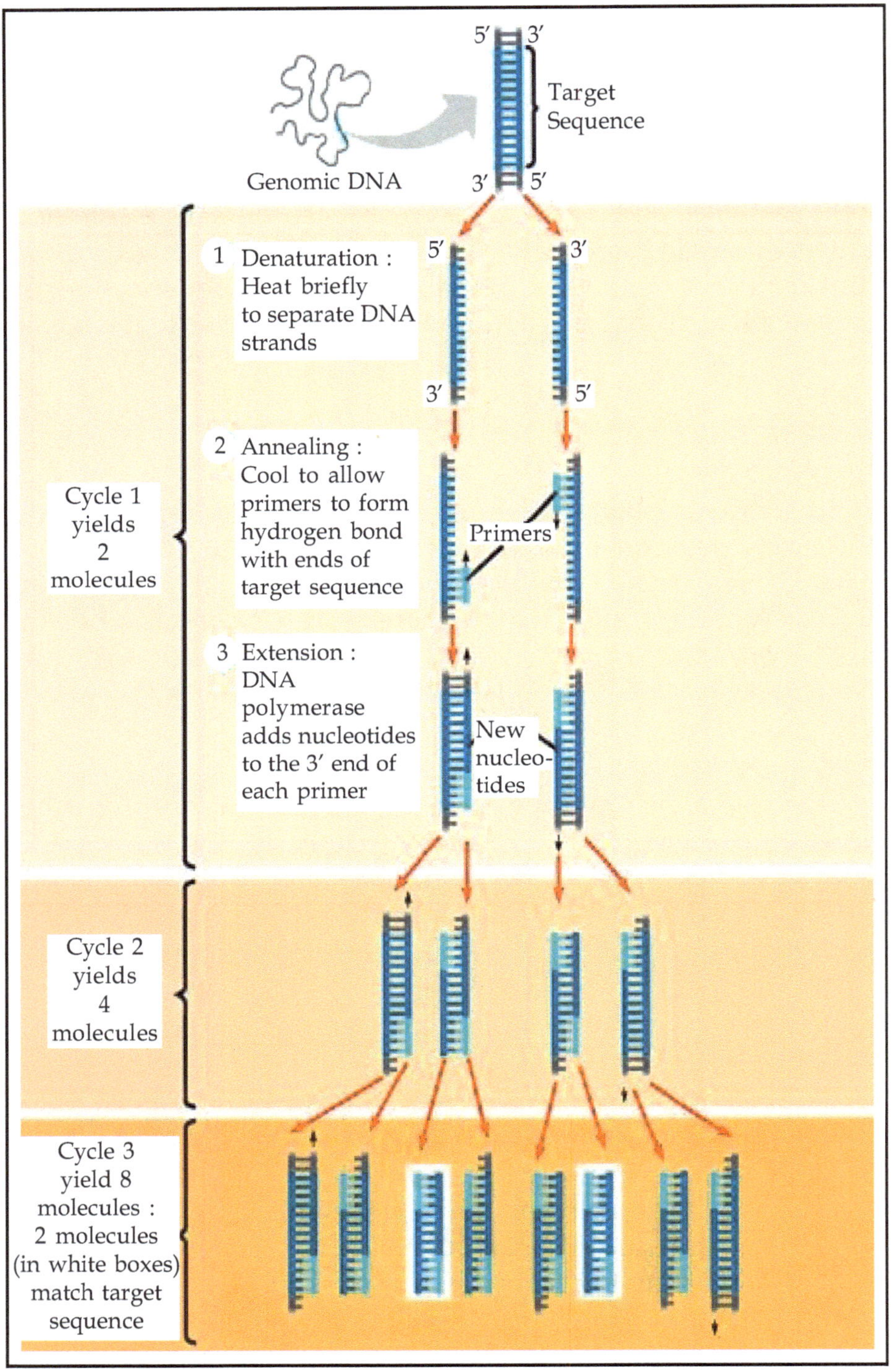

Figure 3: Schematic diagram of Polymerase chain reaction

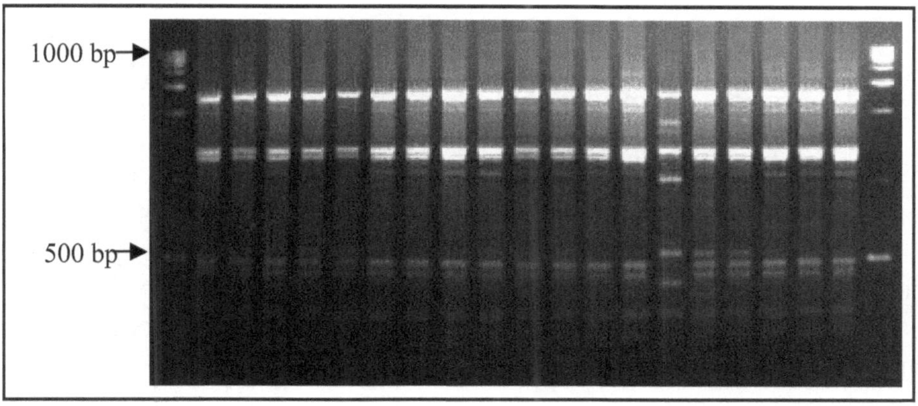

Figure 4: DNA samples amplified with random primers. M-Marker, 1-19 DNA samples

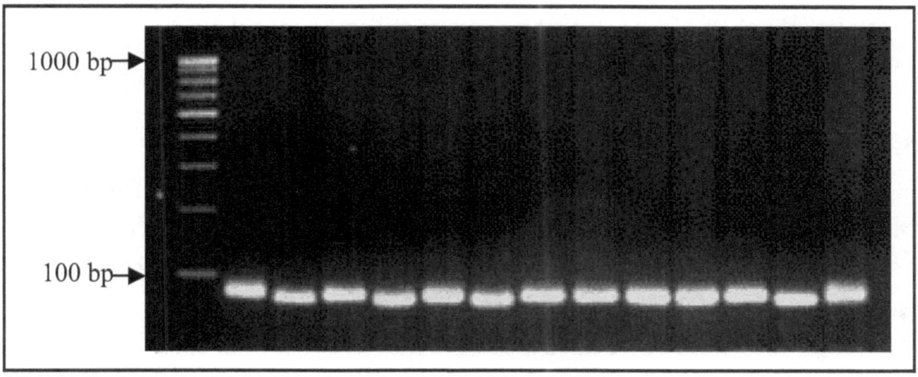

Figure 5: DNA samples amplified with specific primers. M-Marker, 1-13 DNA samples

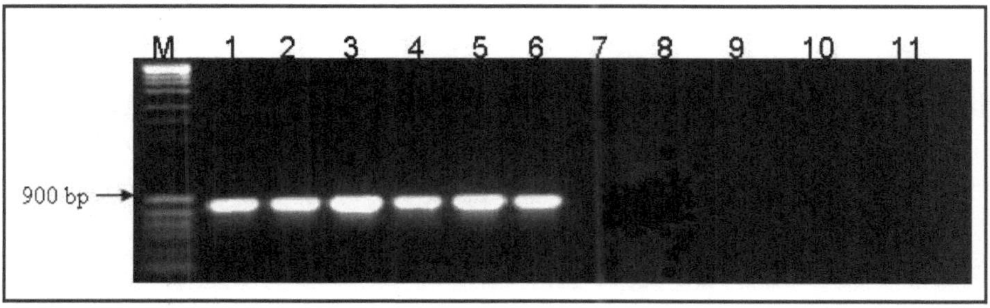

Figure 6: Gel photograph showing PCR products of fts Z Wolbachia specific primers in Uzi flies (*Source:* Dr. Puttaraju, Bangalore University)

Weak product

Gradually decrease the annealing temperature to the lowest possible (touch down PCR). Increase the amount of PCR primer. Increase the amount of DNA template. Increase the amount of Taq polymerase. Change buffer (KCl) concentration (higher if product is lower than 1000bp or lower if product is higher than 1000bp). Add adjuvant and best, use BSA (0.1 to 0.8 μg/μL final concentration). You can also try 5% (v/v, final concentration) DMSO or glycerol. Check primer sequences for mismatches and/ or increase the primer length by 5 nucleotides.

Primers having different melting temperatures

An easy solution is to increase the length of the primer with low Tm. If you need to keep the size of the product constant, add a few bases at the 3' end. If size is not a concern, add a few bases at either the 3' or the 5' end of that primer.

PCR contaminations

Prepare your master mix in a separate room from your current location, somewhere PCR is not the main practice. Be sure to have a separate lab coat, gloves, tubes, pipette tips tube used only in that clean room. Use a separate aliquot of DEPC water stock for each round of PCR. Prepare your mix in a hood with laminar flow. Decontaminate it with bleach, alcohol, RNAase, DNAase, etc... Be sure to UV-irradiate pipettes, pipette tips, tubes, racks, gloves, and also your aliquots of water and PCR buffer before the procedure. Use a different pipette tip when pipetting all your reagents, even the same master mix to each tube. Keep your tubes closed during the procedure, even your master mix tube. Be sure that your tubes are closed when discarding the pipette tip Aerosols are dangerous. Open the tubes only when necessary. More important schedule your PCR when not handling plasmids

Concentration of Mg^{2+}

Mg^{2+} complexes the single nucleotides in your PCR reaction, and only Mg/nucleotide-complexes are the substrate for the DNA-polymerase. So, no Mg^{2+}, no amplification is possible at 10mM

Mg^{2+}, titration means you should try various concentrations of Mg^{2+} (e.g. 1, 3, 5, 7, 9 mM) for the same PCR reaction, since the concentration of Mg^{2+} influences the productivity and fidelity of polymerases. No PCR-product at 1mM Mg^{2+} and only smear at 10mM Mg^{2+} doesn't mean this reaction is not going to work at all.

References

Mullis, K. B. (1990) The Unusual Origin of the Polymerase Chain Reaction. *Scientific American*. 56-65.

Mullis, K.B. and Faloona, F.A. (1987) Specific synthesis of DNA *in vitro* via a polymerase catalyzed chain reaction". *Meth. Enzymol.* 155, 335-350.

Robert F. (1995) The Incredible Shrinking Laboratory. *Science*. 268, 26-27.

❑❑❑

Chapter 7

Electrophoresis

Introduction

Gel electrophoresis is a technique for separating charged molecules with different sizes. Two kinds of gels are commonly used: agarose and polyacrylamide. Agarose gels can be applied to a wider range of sizes than polyacrylamide gels. By using standard agarose electrophoresis, nuclei acids up to 50 kb may be separated. If Pulsed Field Gel Electrophoresis is used, the upper limit can be extended to 10 Mb. Polyacrylamide gels may separate nucleic acids that differ in length by only 1 nucleotide if their length is less than 500 bp (Sambrook *et al.*, 1989).

Principle

Electrophoresis is a process of moving charged molecules in solutions by applying an electrical field across the mixture. Because molecules in an electric field move with a speed dependent on their charge, shape, and size, electrophoresis has been extensively developed for molecular separation. As an analytical tool, electrophoresis is simple and relatively rapid. It is used chiefly for analysis and purification of very large molecules such as proteins and nucleic acids, but can be applied to simpler charged molecules, including charged sugars, amino acids, peptides, nucleotides, and simple ions, highly sensitive detection method have been developed to monitor and analyze electrophoretic separations.

Electrophoresis of macromolecules is normally carried out by applying a thin layer of a sample to a solution stabilized by a porous matrix. Under the influence of an applied voltage, different species of molecules in the sample move through the matrix at different velocities. At the end of the separation, the different species are detected as bands at different positions in the matrix. A matrix is required because the electric current passing through the electrophoresis solution generates heat, which causes diffusion and convective of the bands in the absence of a stabilizing medium.

Agarose Gel Electrophoresis

Background

Agarose gel electrophoresis is the easiest and commonest way of separating and analyzing DNA. The purpose of the gel might be to look at the DNA, to quantify it or to isolate a particular band. The DNA is visualized in the gel by addition of ethidium bromide. This binds strongly to DNA by intercalating between the bases and is fluorescent meaning that it absorbs invisible UV light and transmits the energy as visible orange light.

Buffers and Solutions

Tris Borate EDTA

People also use TAE (Tris Acetate EDTA). Make up a 10x stock using cheap reagents. Do not use expensive 'analytical grade' reagents. Cheap Tris base and boric acid can be bought in bulk.

Recipe for 2 litre of 10xTBE

- 218g Tris base
- 110g Boric acid
- 9.3g EDTA

Dissolve the ingredients in 1.9 litre of distilled water. pH to about 8.3 using NaOH and make up to 2 litre.

Loading dye

The loading dye gives colour and density to the sample to make it easy to load into the wells. Also, the dyes are negatively charged in neutral buffers and thus move in the same direction as the DNA during electrophoresis. This allows you to monitor the progress of the gel. The most common dyes are bromophenol blue and xylene cyanol. Density is provided by glycerol or sucrose.

Typical recipe

- 25mg bromophenol blue or xylene cyanol
- 4g sucrose
- H_2O to 10ml

The exact amount of dye is not important and store at 4°C to avoid mould growing in the sucrose. 10ml of loading buffer will last for years.

Ethidium Bromide

Add 1g of ethidium bromide to 100ml of double distilled water. Stir on a magnetic stirrer for several hours to ensure that the dye has dissolved. Warp water container in aluminium foil or transfer the solution to a dark bottle and store at 4°C in the dark with TOXIC lable on the container.10mg/ml ethidium bromide solution is made up using tablets (to avoid weighing out powder).

What percentage gel?

Most agarose gels are made between 0.7% and 2%. 0.7% gel will show good separation (resolution) of large DNA fragments (5–10kb) and a 2% gel will show good resolution for small fragments (0.2–1kb). Some people go as high as 3% for separating very tiny fragments but a vertical polyacrylamide gel is more appropriate in this case. Low percentage gels are very weak and may break when you try to lift them. High percentage gels are often brittle and do not set evenly.

Which gel tank?

Small mini gels of 8x10cm are very popular and give good photographs. Larger gels are used for applications such as

Southern and Northern blotting. The volume of agarose required for a mini gel is around 30–50ml, for a larger gel it may be 250ml. This method assumes you are making a mini-gel.

How much DNA should we load?

The big question while preparing an analytical gel is to just look at your DNA. Alternatively, you may be preparing a preparative gel to separate a DNA fragment before cutting it out of the gel for further treatment. Either way you want to be able to see the DNA bands under UV light in an ethidium-bromide-stained gel. Typically, a band is easily visible if it contains about 20ng of DNA.

Now consider an example. Suppose you are digesting a plasmid that comprises 3kb of vector and 2kb of insert. You are using EcoRI (a common restriction enzyme) and you expect to see three bands: the linearised vector (3kb), the 5' end of the insert (0.5kb) and the 3' end of the insert (1.5kb). In order to see the smallest band (0.5kb) you want it to contain at least 20ng of DNA. The smallest band is 1/10th the size of the uncut plasmid. Therefore you need to cut 10x20ng that is 200ng of DNA (0.2µg). Then your three bands will contain 120ng, 20ng and 60ng of DNA respectively. All three bands will be clearly visible on the gel and the biggest band will be six times brighter than the smallest band.

Now imagine cutting the same plasmid with BamHI (another popular restriction enzyme) and that BamHI only cuts the plasmid once, to linearise it. If you digest 200ng of DNA in this case then the band will contain 200ng of DNA and will be very bright and will probably be overloaded.

Too much DNA loaded onto a gel is a bad thing. The band appears to run fast (implying that it is smaller than it really is) and in extreme cases can mess up the electrical field for the other bands, making them appear the wrong size also.

Too little DNA is only a problem in that you will not be able to see the smallest bands because they are too faint.

Having said all that, we usually digest and load 2–4μl of the 50μl obtained from a kit miniprep. But you see how it depends on the number and size of the bands expected.

For PCR reactions, it depends on the PCR but in routine applications 10–20μl should be plenty to see the product on the gel.

Which comb?

This depends on the volume of DNA you are loading and the number of samples. Combs with many tiny teeth may hold 10μl. This is no good if you want to load 20μl of restriction digest plus 5μl of loading buffer. When deciding whether a comb has enough teeth, remember that you need to load at least one marker lane, preferably two.

Protocol

Protocol for 50ml gel

1. Weigh out 0.5g of agarose into a 250ml conical flask. Add 50ml of 0.5xTBE, swirl to mix.

   It is good to use a large container, as long as it fits in the microwave, because the agarose boils over easily.

2. Microwave for about 1 minute to dissolve the agarose.

   The agarose solution can boil over very easily so keep checking it. It is good to stop it after 45 seconds and give it a swirl. It can become superheated and NOT boil until you take it out whereupon it boils out all over you hands. So wear gloves and hold it at arms length. You can use a bunsen burner instead of a microwave - just remember to keep watching it.

3. Leave it to cool on the bench for 5 minutes down to about 60°C (just too hot to keep holding in bare hands).

   If you had to boil it for a long time to dissolve the agarose then you may have lost some water to water-vapour. You can weigh the flask before and after heating and add in a little distilled water to make up this lost volume. While the agarose is cooling, prepare the gel tank ready, on a level surface.

4. Add 1μl of ethidium bromide (10mg/ml) and swirl to mix.

   The reason for allowing the agarose to cool a little before this step is to minimise production of ethidium bromide vapour. Ethidium Bromide is mutagenic and should be handled with extreme caution. Dispose of the contaminated tip into a dedicated ethidium bromide waste container.

5. Pour the gel slowly into the tank. Push any bubbles away to the side using a disposable tip. Insert the comb and double check that it is correctly positioned.

   The benefit of pouring slowly is that most bubbles stay up in the flask. Rinse out the flask immediately.

6. Leave to set for at least 30 minutes, preferably 1 hour, with the lid on if possible.

   The gel may look set much sooner but running DNA into a gel too soon can give terrible-looking results with smeary diffuse bands.

7. Pour 0.5x TBE buffer into the gel tank to submerge the gel to 2–5mm depth. This is the running buffer.

   You must use the same buffer at this stage as you used to make the gel. ie. If you used 0.6xTBE in the gel then use 0.6xTBE for the running buffer. Remember to remove the metal gel-formers if your gel tank uses them.

   Preparing the samples

8. Transfer an appropriate amount of each sample to a fresh microfuge tube.

   It may be 10μl of a 50μl PCR reaction or 5μl of a 20μl restriction enzyme digestion. If you are loading the entire 20μl of a 20μl PCR reaction or enzyme digestion then there is no need to use fresh tubes. Write in your lab-book the physical order of the tubes so you can identify the lanes on the gel photograph.

9. Add an appropriate amount of loading buffer into each tube and leave the tip in the tube.

   Add 0.2 volumes of loading buffer, eg. 2μl into a 10μl sample. The tip will be used again to load the gel.

10. Load the first well with marker.
    We store our markers ready-mixed with loading buffer at 4°C. We know to load 2μL and how much DNA is in each band. See below for more on this.

    Avoid using the end wells if possible. For example, If you have 12 samples and 2 markers then you will use 14 lanes in total. If your comb formed 18 wells then you will not be using 4 wells. It is best to not use the outer wells because they are the most likely to run aberrantly.

11. Continue loading the samples and finish of with a final lane of marker.
    We load gels from right to left with the wells facing me. This is because gels are published, by convention, as if the wells were at the top and the DNA had run down the page. If this seems confusing then you can load left to right with the wells facing away from you.

12. Close the gel tank, switch on the power-source and run the gel at 5V/cm.
    For example, if the electrodes are 10cm apart then run the gel at 50V. It is fine to run the gel slower than this but do not run any faster. Above 5V/cm the agarose may heat up and begin to melt with disastrous effects on your gel's resolution. Some people run the gel slowly at first (eg. 2V/cm for 10 minutes) to allow the DNA to move into the gel slowly and evenly, and then speed up the gel later. This may give better resolution.

13. Check that a current is flowing.
    You can check this on the power-source, the milliamps should be in the same ball-park as the voltage, but the best way is to look at the electrodes and check that there are evolving gases (ie. bubbles). If not then check the connections, that the power-source is plugged in etc. This has been known to happen if people use water instead of running buffer.

14. Monitor the progress of the gel by reference to the marker dye.
    Stop the gel when the bromophenol blue has run 3/4 the length of the gel.

15. Switch off and unplug the gel tank and carry the gel (in its holder if possible) to the dark-room to look at on the UV light-box.

    Some gel holders are not UV transparent so you have to carefully place the gel onto the glass surface of the light-box. UV is carcinogenic and must not be allowed to shine on naked skin or eyes. So wear face protection, gloves and long sleeves.

Bromophenol blue migrates at a rate equivalent to 200–400bp DNA. If you want to see fragments anywhere near this size (i.e. anything smaller than 600bp) then uses the other dye because the bromophenol blue will obscure the visibility of the small fragments.

Xylene cyanol migrates at approximately 4kb equivalence. So do not use this if you want to visualize fragments of 4kb.

Size markers

There are lots of different kinds of DNA size markers. In the old days the cheapest defined DNA was from bacteriophage so a lot of markers are phage DNA cut with restriction enzymes. Many of these are still very popular eg, lambda HindIII, lambda PstI, PhiX174 HaeIII. These give bands with known sizes but the sizes are arbitrary. Choose a marker with good resolution for the fragment size you expect to see in you sample lanes. For example, for tiny PCR products you might choose PhiX174 HaeIII but for 6kb fragments you would choose lambda HindIII. More recently, companies have started producing ladder markers with bands at defined intervals, eg. 0.5, 1, 1.5, 2, 2.5kb and so on up to 10kb. If you know the total amount of DNA loaded into a marker lane, and you know the sizes of all the bands, you can calculate the amount of DNA in each band visible on the gel. This can be very useful for quantifying the amount of DNA in your sample bands by comparison with the marker bands. It is good to load two markers lanes, flanking the samples.

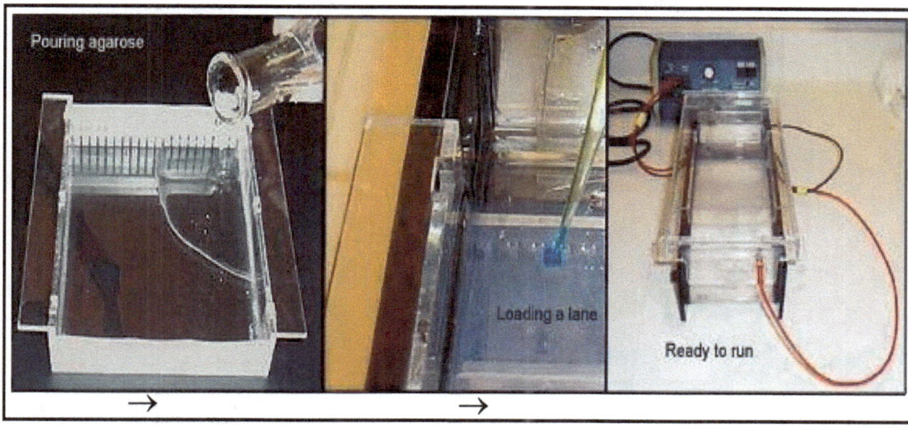

Figure 7: Different stages of agarose gel electrophoresis

Polyacrylamide Gel Electrophoresis

Polyacrylamide (poly (2-propenamide) or poly (1-carbamoylethyelene) is a polymer (-$CH_2CHCONH_2$-) formed from acrylamide subunits that can also be readily cross-linked. In the cross-linked form, it is highly water-absorbent, forming a soft gel used in such applications as polyacrylamide gel electrophoresis. Acrylamide needs to be handled very carefully since it is a neurotoxin using good laboratory practices.

The following protocol is developed for PAGE analysis on Sequi-Gen® GT apparatus (Bio-Rad, USA). The integrated plate chamber (IPC) modules of various sizes could be used depending on the throughput and the resolution needed. We use 38 x 30 cm IPC module for microsatellite analysis.

Solutions for PAGE

i. Bind silane (γ - methacryloxypropyl trimethoxysilane)

ii. Repel silane (Dimethyldichorosilane)

iii. Aceto-Ethanol: 0.5% acetic acid in 95% EtOH: 100 ml

Absolute EtOH	90.25 ml
Glacial acetic acid	5.00 ml
DdWupto	100ml

iv. Acrylamide bis-acrylamide solution (19:1): 40%, 0.5 L

Acrylamide	190g
Bis-acrylamide	10g
ddW up to	500 ml

v. Acrylamide: bis-acrylamide (N,N′ Methylene bis-acrylamide) solution (4%) :1L

TBE(5X)	100ml
Urea	420 g
A: 8(19:1)40% solution	100ml
ddWupto	1000ml

Take 450 ml ddW, add TBE, warm to 50°C, dissolve urea, make up volume to 900 ml and filter; add 100 ml 40% A: B (19:1) last

vi. TEMED (N, N, N′, N′ - Tetraethyl - ethylenediamine) as supplied Ammonium per sulfate (APS) 10% 1ml

APS	100mg
ddW up to	1 ml

Prepare this fresh every time just before use

vii. Gel solution (4% acrylamide): 80 ml

Take 80 ml 4% acrylamide-bis-acrylamide solution

Add 60 µl TEMED, mix well

De-gas using vacuum source or tap-aspirator for ~ 10 min
Add 600 µl of 10% APS solution, mix

viii. Tris-Borate-EDTA (TBE) Buffer (5X): 1L

Tris base	54.0g
EDTA	4.6g
Boric acid	26.6g
ddWupto	1000ml

Store at RT. If any precipitate is observed, discard the buffer
Dilute 10 times to 0.5X with ddW for use

ix. Fix/stop solution: 10% Acetic acid: 1.5 L

Glacial acetic acid	150ml
ddWupto	1500ml

Can be re-used 5 to 6 times; prepare different batches for use at first step and the later steps of gel processing

x. Silver-stain solution: 1.5L

$AgN0_3$	1.5g
Formaldehyde (37%)	2.25 ml
ddWupto	1500ml

Store at RT in dark colored container; May be re-used 2 to 3 times [Add NaCl (5 g/L) to used stain before discarding]

xi. Developer: 1.5 L

Na_2CO3	45.0g
Formaldehyde (37%)	2.0 ml
Sodium thiosulfate (10 mg/ml	300μl
ddWupto	1500ml

xii. STR loading dye: 3X, 10 mL

NaOH (5M)	20 μL
Formamide (95%)	9.5 mL
Bromophenol blue	5 mg
Xylene cyanol FF	5 mg
ddWupto	10 mL

Store at RT.

Protocol

Preparation of notched plate

- Mix 3 μl of bind silane in 1 ml aceto-ethanol
- Dispense on the plate and apply uniformly with tissue paper
- Allow to dry for 5 min
- Remove excess by wiping lightly with tissue paper wetted in 2 ml of 95% EtOH
- If needed, another coat of bind silane may be applied

Preparation of the large plate

- Wipe the plate with absolute EtOH using tissue paper
- Apply and spread 250 µl of repel silane dissolved in 750 µl of the Aceto-ethanol with a tissue paper, wearing hand-gloves
- Wipe the excess with dry tissue paper
- Allow to dry for 5 min

Preparation of gel sandwich

- Place 0.4 mm spacers along the vertical edges of the notched plate
- Place the large plate with coated side facing the spacers taking care to avoid contact between the two coated surfaces
- Prepare 80 ml degassed gel solution (4% acrylamide gel solution)
- Keep the glass plate assembly tilted, and introduce the gel solution steadily from the bottom, till it reaches the top
- Introduce the straight-side of the shark-tooth comb (0.45 mm) in such a way that no air bubble is trapped and the jagged edge of the comb is even with the edge of the large plate
- Allow the gel to polymerise for 20 mm at RT (polymerise gel can be wrapped in ddW soaked paper towels and plastic wrap for storage at RT overnight or by attaching it to the gel apparatus and filling the upper and lower buffer chambers with dry paper towel
- Fix the gel-sandwich to the gel apparatus with 0.5X TBE in the lower buffer chamber
- Remove the comb carefully
- Add 0.5X TBE to the upper buffer chamber
- Rinse the well surface thoroughly with the buffer, using large volume pipette

Pre-run

- Pre-run the gel at 100 W, 2000 V maximum, 50 mA setting (for about 45 min to Ih) to achieve a gel surface temperature of about 45-50 °C

Sample Preparation

- For each μl of sample taken, add 0.5 μl of 3X STR loading dye
- Denature the samples at 95°C for 5 min (preferably using a thermocycler)
- Chill the samples immediately by placing them on ice (denatured samples can be stored at -20°C for a week)
- Prepare the DNA size markers also the same way (50 ng/ lane)

Loading the gel

- After the pre-run, rinse the well area clean of crystallized urea with a high volume pipette, taking care not to disturb the flat well surface
- Carefully insert the shark tooth comb (0.45 mm) with the jagged surface into the well area till the point the tips of the teeth just touch the surface of the gel (loose fitting or sloppy placement of the comb would result in mixing of samples loaded) Quick spin the denatured samples just before loading the gel
- Load 4 μl denatured samples per lane, completing the loading within 20 min to prevent the gel temperature falling below 45 °C

Gel run

- Run the gel at 100 W, 2000V maximum, 50 mA with a temperature bound of 50°C (optimal temperature is 48°C)
- Gel run may be interrupted (not beyond 20 min each time) for multiplex loading (but every time samples are to be loaded, the wells have to be rinsed to get rid of use urea crystallised in them during the run)

- ❐ Put off the power when the xylene-cyanol dye front has run about 2/3 of the gel length (On a 4% PAGE, the bromophenol blue dye runs along 170 b DNA fragments)

Removing gel

- ❐ Drain the buffer from the upper and lower buffer chambers; remove the sandwich from the apparatus

Gel processing

- ❐ Place the large plate with gel surface up in a tray
- ❐ Sequentially follow the steps below with continuous shaking (about 1.5L solution for each step would be needed):
    - Fix in Fix/stop solution for 20 min or till the loading dyes disappear (maximum of 1 h)
    - Rinse in ddW for 2 min, 3 times
    - Stain in the Stain solution for 25 min (minimum 20; maximum 30 min)
    - Rinse in ddW for 5 sec (exceeding 5 sec would reduce the band intensity, through re-staining the gel is possible)
    - Develop in the chilled Developer (4 ~ 10°C) for 2 to 5 min (keep watching for the satisfactory intensity of bands and stop developing when desired intensity is achieved; low temperature developing helps in more accurate control over the developing process)
    - Stop development by rinsing in the Fix/stop solution for 5~-6 min
    - Wash in ddW for 2-3 min, 3 times
    - Air dry the gel overnight
    - Scan the gel in a scanner and acquire the image into a computer with image processing software.

Cleaning the gel plates for re-use

Notched plate is rinsed in warm water, de-ionized water and air-dried in dust free atmosphere. Long glass plate with the dried gel is soaked in NaOH (0.1% w/v overnight or 10% w/v for 30 min) to free from polyacrylamide (may need scraping with blade) Rinse in detergent, tap water, de-ionized water, ddW and air dry in dust-free atmosphere.

Effective range of separation of Polyacrylamide gels

Acrylamide* concentration (%)	Linear range of separation (kD)
15	12-43
10	16-68
7.5	36-94
5.0	57-212

* Molar ratio of Bisacrylamide: Acrylamide is 1:29

NB: *Acrylamide Monomer is A Potent Cummulative Neurotoxin. Do not Mouth Pipette Acrylamide Solutions, and Wear Gloves when Handling Unpolymerised Solutions.*

Applications

- Gel electrophoresis is the main means for separation of nucleic acids on solid media.
- Gel electrophoresis through agarose gels and PAGE can be used effectively for the separation of different sizes of DNA, RNA and proteins fragments.

Troubleshooting

1. Make sure that electrode strips are correctly placed
2. Check polarity: check pH of electrode strips
3. Incorrect electrode solution; use recommended solution
4. Load optimum concentration of proteins/ nucleic acid
5. Give optimum staining time
6. Give optimum (standardized) voltage

7. Optimize the concentration of sample per well
8. Optimize the stain concentration
9. Thoroughly clean gel equipments like tank, combs, plates, spacers, etc.,
10. Use even surface for preparation of gels
11. Use gloves when handling gels.

References

Brody, J. R., Calhoun, E. S., Gallmeier, E., Creavalle, T. D & Kern S. E. (2004) Ultra-fast high-resolution agarose electrophoresis of DNA and RNA using low-molarity conductive media. *BioTechniques* 37:598-602.

Sambrook J, Fritsch EF, Maniatis T (1989) *Molecular cloning: a laboratory manual. Second edition*. Cold Spring Harbor Laboratory Press, New York, USA.

□□□

Chapter 8

Restriction Digestion of Enzymes

Introduction

Restriction enzymes, recognizes specific base sequences in double-helical DNA and cleaves at specific places, both strands of a duplex containing the recognized sequences. To biochemists, these exquisitely precise scalpels are marvelous gifts of nature. They are indispensable for analyzing chromosome structure, sequencing very long DNA molecules, isolating genes, and creating new DNA molecules that can be cloned. Werner Arber and Hamilton Smith discovered restriction enzymes, and Daniel Nathans pioneered their use in the late 1960s.

Principle

Many restriction enzymes recognize specific sequences of four to eight base pairs and hydrolyze a phosphodiester bond in each strand in this region. A striking characteristic of these cleavage sites is that they almost ways possess twofold rotational symmetry. In other words, the recognized sequence is palindromic, or an inverted repeat and the cleavage sites are symmetrically positioned.

More than 100 restriction enzymes have been purified and characterized. Their names consist of a three-letter abbreviation for the host organism (e.g., Eco for *Escherichia coli*, Hin for *Haemophilus influenzae*, Hae for *Haemophilus aegyptius*) followed by a strain designation (if needed) and a roman numeral (if more than one restriction enzyme from the same strain has been identified). Restriction enzymes are used to cleave DNA molecules into specific fragments that are more readily analyzed and manipulated than the entire parent molecule.

Types of Restriction Enzymes

Type I

Type I restriction systems consist of a single enzyme that performs both modification methylation and restriction activities. These enzymes recognize specific DNA sequences, but cleave the DNA strand randomly, at least 1,000 base pairs (bp) away from the recognition site.

Type II

Type II restriction enzymes; in contrast, are heavily used in recombinant DNA techniques. They consist of single, separate proteins for restriction and modification. One enzyme recognizes and cuts DNA; the other enzyme recognizes and methylates the DNA. They cleave the DNA sequence at the same site at which they recognize it. The only exception is type II (shifted) restriction enzymes, which cleave DNA on one side of the recognition sequence, within twenty nucleotides of the recognition site. The enzymes discovered to date collectively recognizes over 200 different DNA sequences.

Type II restriction enzymes can cleave DNA in one of three possible ways. In one case, these enzymes cleave both DNA strands in the middle of a recognition sequence, generating blunt ends. *For example:* The notations 5' and 3' are used to indicate the orientation of a DNA molecule. The numbers 5 and 3 refer to specific carbon atoms in the deoxyribose sugar in DNA.

Type III

Type III restriction systems have separate enzymes for restriction and methylation, but these enzymes share a common subunit. These enzymes recognize specific DNA sequences, but cleave DNA at random sequences approximately twenty-five bp from the recognition sequence.

Type IV

The restriction enzymes target only methylated DNA.

Types of Cuts

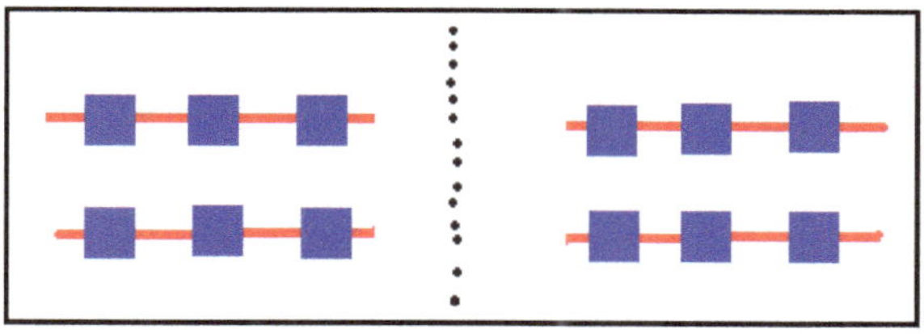

Blunt end Cutters

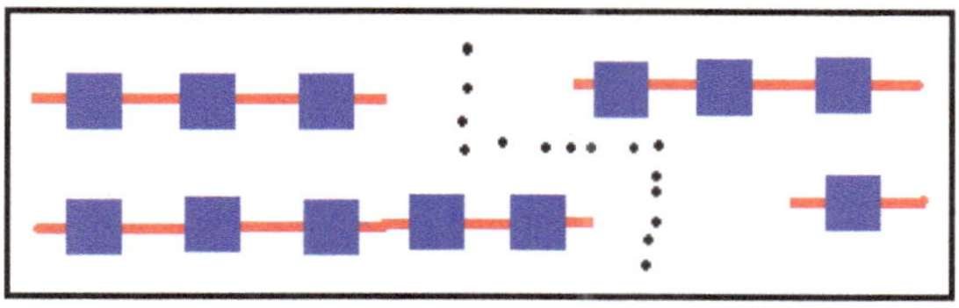

Sticky-end Over-hangers

Restriction digests

Always cut a lot of your starting plasmids in a small volume; this will help in the gel purification of your restriction fragments by giving you a high concentration of DNA compared to agarose in your gel slice. About 1-2 μg in a 20 μl reaction is good.

Restriction enzyme digestions are performed by incubating double-stranded DNA molecules with an appropriate amount of restriction enzyme, in its respective buffer as recommended by the supplier, and at the optimal temperature for that specific enzyme. The optimal sodium chloride concentration in the reaction varies for different enzymes, and a set of three standard buffers containing three concentrations of sodium chloride are prepared and used when necessary. Typical digestions included a unit of enzyme per microgram of starting DNA, and one enzyme unit usually (depending on the supplier) is defined as the amount of enzyme needed to completely digest one microgram of double-stranded DNA in one hour at the appropriate temperature. These reactions usually are incubated for 1-3 hours, to insure complete digestion, at the optimal temperature for enzyme activity, typically 37°C.

Protocol

1. Prepare the reaction for restriction digestion by adding the following reagents in the order listed to a micro centrifuge tube:

sterile ddH20	q.s (where "q.s." means quantity sufficient)
10X assay buffer	one-tenth volume
DNA	x µl (0.5-2 µg)
Restriction enzyme*	y µl (1-10 units per µg DNA)
Total volume	z µl

**If desired, more than one enzyme can be included in the digest if both enzymes are active in the same buffer and the same incubation temperature.*

NOTE: The volume of the reaction depends on the amount and size of the DNA being digested.

2. Gently mix by pipetting and incubate the reaction at the appropriate temperature (typically 37° C) for 1-3 hours.
3. Inactivate the enzyme(s) by heating at 70-100 °C for 10 minutes and store at -20 °C.
4. Gel purifying DNA fragments: Purify all DNA fragments using gel purification kit.

When the gel is done, place it on a clean piece of saran wrap on a long wave (365 nm) UV box and take a picture. Make sure minimize the exposure to UV to avoid damaging the DNA. Cut the desired bands out of the gel with a clean scalpel. Place the gel slice in an eppendorf tube. Freeze the gel slice to store.

Applications

Recombinant DNA Technology

Perhaps the most important use of restriction enzymes has been in the generation of recombinant DNA molecules, which are DNAs that consist of genes or DNA fragments from two different organisms. Typically, bacterial DNA in the form of a plasmid (a small, circular DNA molecule) is joined to another piece of DNA (a gene) from another organism of interest. Restriction enzymes are used at several points in this process. They are used to digest the DNA from the experimental organism, in order to prepare the DNA for cloning. Then a bacterial plasmid or bacterial virus is digested with an enzyme that yields compatible ends. These compatible ends could be blunt (no overhang), or have complementary overhanging sequences. DNA from the experimental organism is mixed with DNA from the plasmid or virus, and the DNAs are joined with an enzyme called DNA ligase.

Forensic DNA Fingerprinting and Southern Blotting

Restriction fragment length polymorphism (RFLP) analysis has been widely used for identification of individuals (humans and other species). In this technique, genomic DNA is isolated, digested with a restriction enzyme, separated by size in an agarose gel, and then transferred to a membrane. The digested DNA on the membrane is allowed to bind to a radioactively or fluorescently labeled probe that targets specific sequences that are bracketed by restriction enzyme sites. The size of these fragments varies in different individuals, generating a "biological bar code" of restriction enzyme-digested DNA fragments, a pattern that is unique to each individual.

Cloning Verification

As the restriction enzymes cleave the DNA fragments at a particular site the modified enzymes to cut a particular DNA sequence can be used for verification of cloned materials.

Troubleshooting

Buffer Composition

Different restriction enzymes have differing preferences for ionic strength (salt concentration) and major cat ion (sodium or potassium). A battery of 3 to 4 different buffers will handle a large number of available enzymes, although there are a few that require a unique buffer environment. In all cases, a major function of the buffer is to maintain pH of the reaction (usually at 8.0). Additionally, some enzymes are fussier about having their optimal buffer than other enzymes. Clearly, use of the wrong buffer can lead to poor cleavage rates.

Incubation Temperature

Most restriction enzymes cut best at 37°C, but there are many exceptions. Enzymes isolated from thermophilic bacteria cut best at temperatures ranging from 50 to 65°C. Some other enzymes have a very short half life at 37°C and it is recommended that they be incubated at 25°C.

Influence of DNA Methylation

Almost all strains of *E. coli* bacteria used for propagating cloned DNA contain two site-specific DNA methylases:

- *Dam methylase* adds a methyl group to the adenine in the sequence GATC, yielding a sequence symbolized as G^{m}ATC.
- *Dcm methylase* methylates the internal cytosine in CC (A/T) GG, generating the sequence C^{m}C (A/T) GG.

Digestion with Multiple Enzymes

Digesting DNA with two enzymes is a commonplace task, and oftentimes the two enzymes have different buffer requirements. There are at least three ways to handle this situation:

- Digest with both enzymes in the same buffer. In many cases, even those given buffer is not optimal for an enzyme, you can still get quite good cleavage rates. Enzyme manufacturer catalogs usually contain a reference table recommended the best single buffer for conducting specific double digests.
- Cut with one enzyme, then alter the buffer composition and cut with the second enzyme. This usually applies to situations where one enzyme like a low salt buffer and the other a high salt buffer, in which case you can digest with the first enzyme for a time, add a calculated amount of concentrated NaCl and cut with the second enzyme.
- Change buffers between digestions with two enzymes. In some cases, two enzymes will have totally incompatible buffers. In that case, perform one digestion, recover the DNA (usually by precipitation) and resuspend in the buffer appropriate for the second enzyme.

Variability in Digestion of Different DNA Substrates

The efficiency with which a restriction enzyme cuts its recognition sequence at different locations in a piece of DNA can vary 10 to 50-fold. This is apparently due to influences of sequences bordering the recognition site, which perhaps can either enhance or inhibit enzyme binding or activity.

A related situation is seen when restriction recognition sites are located at or very close to the ends of linear fragments of DNA. Most enzymes require a few bases on either side of their recognition site in order to bind and cleave. Many of the companies that sell enzymes provide a table in their catalog that presents "end requirements" for a variety of enzymes.

References

Bloom, Mark V., Greg A. F & David A. M. (1996) *Laboratory DNA Science: An Introduction to Recombinant DNA Techniques and Methods of Genome Analysis.* Menlo Park, CA: Addison-Wesley.

Cooper, Geoffrey (1997) *The Cell: A Molecular Approach.* Washington, DC: ASM Press.

Kreuzer, Helen, & Adrianne Massey. (2000) *Recombinant DNA and Biotechnology,* 2nd ed. Washington, DC: ASM Press.

Lodish, H. *Molecular Cell Biology,* 4th ed. New York: W. H. Freeman, 2000.

Old, R. W. & Primrose S. B. (1994) *Principles of Gene Manipulation,* 5th ed. London: Blackwell Scientific Publications.

❑❑❑

Chapter 9

Blotting

Southern Blotting

Introduction

Southern hybridization was named after Edward M. Southern who developed this procedure at Edinburgh University in the 1970s. To oversimplify, DNA molecules are transferred from an agarose gel onto a membrane. Southern blotting is designed to locate a particular sequence of DNA within a complex mixture. It is also called Southern blotting since the procedure for transfer of DNA from the gel to the nitrocellulose filter resembles blotting. In Southern hybridization, a sample of DNA containing fragments of different sizes is subjected to electrophoresis using either polyacrylamide or agarose gel. The DNA sample may either be subjected to mechanical shearing or to restriction endonuclease digestion in order to generate the fragments. Agarose gel is useful in separating DNA fragments of few hundred to 20 kb in size, while polyacrylamide is preferred for smaller fragments. Very large DNA fragments up to 1000-2000 kb are separated in agarose gel with pulsed electrical fields or field inversion.

Principle

The procedure starts with the restriction enzyme digestion of genomic DNA, followed by agarose gel electrophoresis and Southern Blotting. The DNA used in this procedure is isolated according to protocol DNA isolation from blood or DNA isolation from chorionic villi. Five μg of each DNA-sample is digested by one or more restriction enzymes. After incubation all samples are tested for complete digestion. If the samples are digested completely, the procedure can be continued; otherwise an extra incubation step is performed.

The procedure continues with the separation of the DNA fragments on an agarose gel. An electric field is applied across the agarose gel on which the DNA samples are loaded. At neutral pH the DNA-fragments have a negative charge and will therefore migrate towards the anode (positively charged electrode). The DNA-fragments are separated according to their size; the smaller fragments will migrate faster through the gel than the larger fragments.

After separation the DNA-fragments are transferred to a nylon membrane. The DNA in the gel is denatured with alkali-solution; subsequent capillary transport of the DNA fragments takes place from the gel to the nylon membrane, using the same alkali-solution. The relative positions of the DNA-fragments are preserved during their transfer to the nylon membrane.

After alkali-blotting the DNA-fragments are bound to the membrane and hybridization with a radio-labelled probe is used to detect specific fragments.

What Southern blotting will tell us?

1. Whether a particular gene is present and how many copies are present in the genome of an organism.
2. The degree of similarity between the chromosomal gene and the probe sequence.
3. Whether recognition sites for particular restriction endonucleases are present in the gene. By performing the digestion with different endonuclease, or with combinations

of endonucleases, it is possible to obtain a restriction map of the gene i.e. an idea of the restriction enzyme sites in and around the gene- which will assist in attempts to clone the gene.

4. Whether re-arrangements have occurred during the cloning process.

Reagents and equipments

- Balance
- Combs, 24 or 28 teeth, 1.0 mm thick
- Combs, 22 teeth
- Ultra Pure Water System
- Freezer
- Glass plates (15x19 cm, 19x19 cm and 20x26 cm)
- Glassware (M)
- Horizontal electrophoresis system
- Icubator
- Magnetic stirrer
- Micro centrifuge
- Microwave Oven
- Minishaker
- Pipetman P20, P200, P1000
- Power supply (Powerpac 300)
- Refrigerator
- Sub-Cell GT Agarose Gel Electrophoresis Systems
- The Imager

NOTE: *Equivalent durables may be used*

Consumables

- Blue Tip
- Cell Saver Tips

- Filter paper
- GB002 Filter paper
- GB003 Filter paper
- Hybond XL membrane
- Jif
- Micro-titre plate
- Paper towels
- Plastic trays (20x16 cm)
- Safelock Tubes (1.5 ml, Eppendorf)
- Thermal printer paper glossy
- Tip loading refill

NOTE: *Equivalent consumables may be used*

Chemicals

- Bromophenolblue (sodium salt) 10 g
- Ethanol (EtOH) 5 litre (96 %), 01-0181
- Hydrochloric acid (HCl) 1 litre (36-38 %)
- Multipurpose agarose 500 g
- Sodium Dodecyl Sulfate (SDS) 1 kg
- Sodium chloride (NaCl) 1 kg
- Sodium hydroxide (NaOH) 1 kg
- Spermidine 1 g
- Sucrose 1 kg

NOTE: *Equivalent consumables may be used*

Solutions

The solutions are prepared according to the protocols and autoclaved when necessary. For the solutions marked with a * a special form (with the protocol on it) is filled out. The other solutions are ready to use or made according to the protocols described here.

- *Alkali-solution 1 (0.4 N NaOH);* take 80 g NaOH, add pure water to a total volume of 5 litre and dissolve on a magnetic stirrer for at least 15 min.

  NOTE: *Prepare fresh before use. For one gel prepare half the amount*

- *Alkali-solution 2 (0.4 N NaOH, 0.6 M NaCl);* take 80 g NaOH and 175.3 g NaCl, add Elga water to a total volume of 5 litre and dissolve on a magnetic stirrer for at least 15 min.

  NOTE: *Prepare fresh before use. For one gel prepare half the amount*

- *Ethanol (EtOH) 70 %;* add 150 ml Elga water to 350 ml technical ethanol (96 %).
- *Ethidium bromide (EtBr);* 10 mg/ml.
- *Ethylene diamine tetra acetate (EDTA);* 0.5 M pH8.0.
- *Restriction enzymes*
- *Restriction enzyme reaction buffer;* 10 x
- *Sodium dodecyl sulfate (SDS);* 10 %
- *Sucrose loading mix (SLM-buffer);* 10 x
- 2 x SLM for 100 ml: Dilute 20 ml of 10 x SLM to 100 ml with water. Store at room temperature.
- *Spermidine;* 1.0 M spermidine.
- *SSC;* 20 x SSC
- *Neutralisation buffer;* (2 x SSC, 0.2 M Tris-HCl, pH7.5) for 2 litre add 200 ml 20 x SSC and 400 ml 1 M Tris-HCl (pH7.5) to 1400 ml of water

  NOTE: *The solution can be stored for 1 month at room temperature*

- *TAE-buffer;* 10 x Tris Acetate EDTA buffer stock

  NOTE: *Store at room temperature*

- *TAE/EtBr buffer;* 0.4 mg EtBr/l TAE-buffer

  NOTE: *Store at room temperature*

- *TBE-buffer;* 10 x Tris Borate EDTA buffer stock

  NOTE: *Store at room temperature*
- *TBE/EtBr buffer;* 0.2 mg EtBr/l TBE-buffer

  NOTE: *Store at room temperature*
- *Tris-HCl (pH7.5);* 1 M

Protocol

1. Digestion of DNA sample

Protocol below is based on a digestion of *5 µg of genomic DNA,* which is sufficient for one lane on the blot gel. If necessary it is possible to take a multiple amount of DNA. The complete procedure takes 3 days. For pipetting water, spermidine and buffer there is no need to change tips.

NOTE: *DNA isolated from chorionic villi contains a lot of RNA. This influences the measured DNA concentration. To get equal signals on your blot compared to normal blood-derived DNA samples, take twice the amount for the digestion*

1. Based on the measured DNA-concentration, calculate which volume has to be pipetted from each DNA sample.
2. Write down the amount of water (in µl) you have to add, to get a total volume of 23 µl.

   NOTE: in some cases it is necessary to adjust this volume
3. Mark the 1.5 ml Eppendorf safe lock tubes with the complete DNA number and the code(s) of the enzyme(s) used.
4. Prepare a mix sufficient for all digestions, containing 3 µl of the 10 x restriction enzyme reaction buffer mentioned in the disease specific protocol and 1 µl 100 mM spermidine.

   NOTE: *In some cases it is necessary to adjust the amount of 10 x restriction enzyme reaction buffer used.*
5. First pipette the amount of water in each tube, then add 4 µl of the restriction enzyme reaction buffer/spermidine mix.
6. Using Cell Saver tips, add the calculated volume of each DNA sample.

NOTE: *Use a new tip for every DNA sample. The total volume is 27 μl, or a multiple thereof*

7. Incubate 30 min in the 65°C incubator, cool down to room temperature and add 15 units restriction enzyme (3 units for each μg of DNA).

   NOTE: *Never add more enzyme than 10 % of the total volume because glycerol inhibits enzyme activity. Mix carefully and spin down 5-10 sec in the Eppendorf centrifuge*

8. Incubate overnight in the 37°C incubator.

   NOTE: *For some diseases enzymes with different incubation temperatures are used, or double digestions*

2. Analysis of the digestion

This part of the protocol is partly carried out in a pre-PCR lab (preparation of the samples) and partly in a post-PCR lab (preparation of the gel and electrophoresis of the samples). Depending on the capacity needed, two gel sizes are used: a 15 x 45 cm gel (capacity 6 x 22 lanes) or a 15 x 10 cm gel (capacity 2 x 22 lanes). For more information see electrophoresis chapter.

Blotting

STEP 1

NOTE: *Alkali-solution 1 is used to transfer the DNA in 4 to 6 hrs, alkali-solution 2 is used for overnight blotting*

1. For every blot, prepare a fresh alkali-solution.

2. Clean all the glass plates and trays with water and, if necessary, with "Jif".

3. Put the gel on the UV transilluminator.

   NOTE: *Clean the UV transilluminator with water before and after putting the gel on the transilluminator*

4. Cut the gel to the correct size (just above the slots, and just around the DNA).

5. Make a print-out of the gel with the Imager system (use the print button on the computer, date and time will be printed).
6. Stick the print-out on the WBLOT-form.
7. Turn the gel upside down between two glass plates.
8. Denature the gel two times 15 min in the alkali-solution under gentle agitation (number the gels when more then one gel is used in the blotting procedure) Refresh the solution after 15 min.

   NOTE: *In the mean time prepare everything needed for blotting*

   - ❒ Cut the following to size
   - ❒ Two GB002 papers (35 x 19 cm or 46 x 19 cm)
   - ❒ One GB002 paper (20 x 19 cm or 29 x 19 cm)
   - ❒ One Hybond XL membrane, 1 cm larger than the gel. Mark the membrane with a black pen with the date and number of the gel (if more then one gel is blotted). Put water in a tray, soak the membrane in it, pour off the water and leave the membrane in alkali-solution for at least 5 min
   - ❒ One GB002 paper same size as the gel
   - ❒ One GB003 paper same size as the gel

STEP 2

1. Start to build the "blotting tower"; put a glass plate on top of 2 plastic trays (20 x 16 cm) containing alkali-solution.
2. Soak the two GB002 papers (35 x 19 cm or 46 x 19 cm) in alkali-solution and put them on the glass plate. Remove air bubbles by rolling with a glass pipette.

   NOTE: Repeat this step every time you put a new layer on the blotting tower

3. Soak the GB002 paper (20 x 19 cm or 29 x 19 cm) in the alkali-solution and put it on the other papers.

4. Add alkali-solution to soak the paper bridge completely.
5. Put the denatured gel, still upside down, on a glass plate and carefully transfer it to the GB002 paper bridge.

   NOTE: *Be extra careful with a 0.5 % gel*
6. Put slices of film around and just under the edges of the gel.

   NOTE: *This ensures a capillary flow through the gel to the membrane*
7. Add alkali-solution to soak the gel completely.
8. With the marked corner in the correct position, put the membrane on the gel.

   NOTE: *See the disease specific protocols for the exact positions of the marks*
9. Soak the GB002 and the GB003 papers (with the same size as the gel) together in alkali-solution and put them on the membrane.
10. Fold paper towels in about the same size as the gel. Make a pile of about 10 cm high and put this on the blotting tower.
11. Place a glass plate together with a weight of about 250 gram on top.

    NOTE: *Do not use a weight when a 0.5 % agarose gel is blotted*
12. Blotting takes place in 4 to 6 h (with alkali-solution 1) or overnight (with alkali-solution 2).
13. After blotting remove the weight, glass plate, paper towels and papers. Neutralize the membrane in neutralization buffer for 1-5 min and dry it between filter paper.
14. Check the membrane for correct blotting and possible introduction of air bubbles using a UV transilluminator (cover it with Saran Wrap to protect the membrane). Mark all empty lanes and if possible the slots on the DNA side of the membrane. Check this with the print-out of the blotgel before cutting the membrane.

15. Cut the membrane to the correct size for hybridization and write, with a black pen on the DNA side, on each piece the WBLOT number, the enzyme(s) used and the DNA numbers.

16. Membranes are now ready for hybridization.

    NOTE: *Store the dried membranes at RT in envelopes on WBLOT number*

17. Throw away remaining alkali solutions, the blotted gel and all paper used in the procedure. Rinse the glass plates and trays with water to remove all alkali solution. Clean the table with water.

Applications

Restriction Fragment Length Polymorphism (RFLP)

Restriction fragment length polymorphism (RFLP) analysis was one of the first techniques to be widely used for detecting variation at the DNA sequence level. The principle behind the technology rests on the possibility of comparing band profiles generated after restriction enzyme digestion in DNA molecules of different individuals. Diverse mutations that might have occurred affect DNA molecules in different ways, producing fragments of variable lengths. These differences in fragment lengths can be seen after gel electrophoresis, hybridization and visualization.

Variable number tandem repeats (VNTR)

VNTR or minisatellites were suggested by Alec Jeffreys and his colleagues from U.K for the study of DNA polymorphism in humans. Primers were developed for the conserved flanking regions of VNTR loci, so that all available VNTR loci would be amplified. The PCR products differ according to number of repeat units present in the VNTR loci and this variation is visualized positions occupied by different bands after electrophoresis.

Gene cloning

Cloning is the process of recreating an identical copy of an original organism or thing. A cloning in the biological sense, therefore, is a molecule, single cell (like bacteria, lymphocytes etc.) or multi-cellular organism that has been directly copied from and is therefore genetically identical to another living organism. Sometimes this term can refer to "natural" clones made either when an organism reproduces asexually or when two genetically identical individuals are produced by chance (as with identical twins), but in common parlance, a clone is an identical copy created intentionally.

Genome mapping

It can be used to map the restriction sites around a single copy gene sequence in any genome (even of man). It is used for DNA fingerprinting, preparation of RFLP maps, detection and identification of the transferred genes in transgenic individuals etc.

Chromosome walking

When a probe is used for the identification of a gene sequence in a genomic library, the probe may hybridize with a number of clones, each carrying a part of large gene fragmented during preparation of genome library. If we obtain partial digests (by digesting the DNA only partially) from the genome, different genomes (from large number of cells) may give fragments which have overlapping sequences, because sites cleaved in different genomes of the same organism, will differ being random. Since none of these fragments may have entire sequence represented in the probe, overlapping sequences may be used to construct the original genomic sequence. Identification of fragment with overlapping sequences may be a key to the reconstruction or characterization of large chromosome regions. This is achieved by the technique popularly described as Chromosome walking.

Troubleshooting

1. Poor signal

- Probe specific activity too low
- Inadequate depurination
- Inadequate transfer buffer

- Not enough target DNA
- Transfer time too short
- Inefficient transfer system
- Probe concentration too low
- Incomplete denaturation of probe and/or target DNA
- Final wash too stringent
- Hybridization time too short
- Inappropriate membrane

2. Spotty Background
- Unincorporated nucleotides not removed from labeled probe
- Particles in hybridization buffer
- Agarose dried on membrane
- Baking or UV cross linking when membrane contains high salt

3. High Background
- Insufficient Blocking
- Membrane allowing to dry out during hybridization or washing
- Membranes adhered during hybridization or washing
- Bubbles in hybridization bag
- Walls of hybridization bag collapsed on the membrane
- Not enough wash solution
- Hybridization temperature too low
- Labeled probe molecules are too short
- Probe Concentration too high
- Inadequate prehybridization
- Probe not denatured
- Not enough SDS in wash solution

4. When some of the samples are not completely digested, add more enzymes (10-20 units) and incubate again for 3-4 hours at the appropriate temperature. The addition of 1 μl 100 mM

spermidine can also make a difference. If problems remain unsolved perform "Troubleshooting 2"

5. When digestions of a certain DNA sample are all incomplete or remain incomplete, it might be useful to perform an extra purification step according to. This is necessary to get rid of proteins or other contaminants disturbing the digestion.

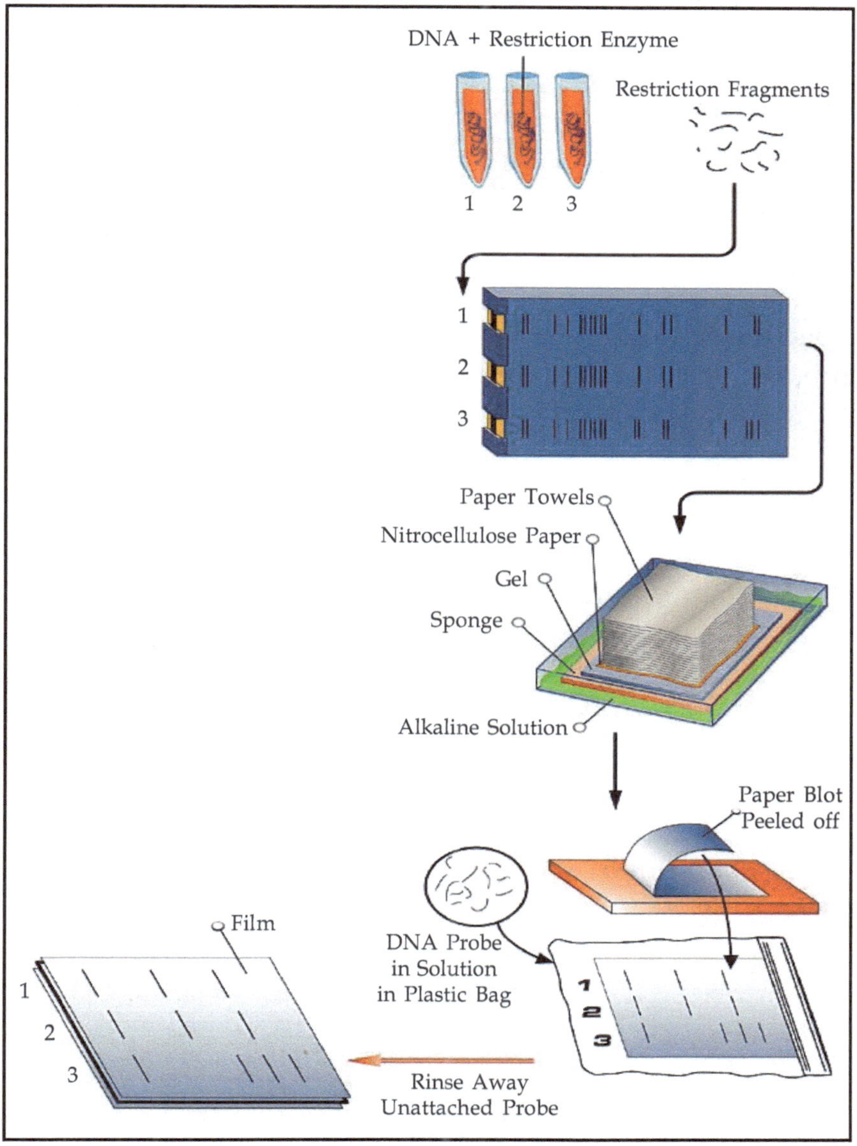

Figure 8: Schematic diagram of Southern blotting

Western Blotting

Introduction

Western blot is a method in molecular biology/biochemistry /immunogenetics to detect protein in a given sample of tissue homogenate or extract. It uses gel electrophoresis to separate denatured proteins by mass. The proteins are then transferred out of the gel and onto a membrane, where they are "probed" using antibodies specific to the protein. As a result, researchers can examine the amount of protein in a given sample and compare levels between several groups. Other techniques also using antibodies allow detection of proteins in tissues and cells. Northern analysis remains a standard method for detection and quantization of mRNA levels despite the advent of powerful techniques, such as RT-PCR, gene array analysis and nuclease protection assays. Northern analysis provides a direct relative comparison of message abundance between samples on a single membrane. It is the preferred method for determining transcript size and for detecting alternatively spliced transcripts.

What is a blot?

Blotting is the process of identifying specific bands of interest amongst all the bands bound to the nitrocellulose. Done with specific probes or antibodies.

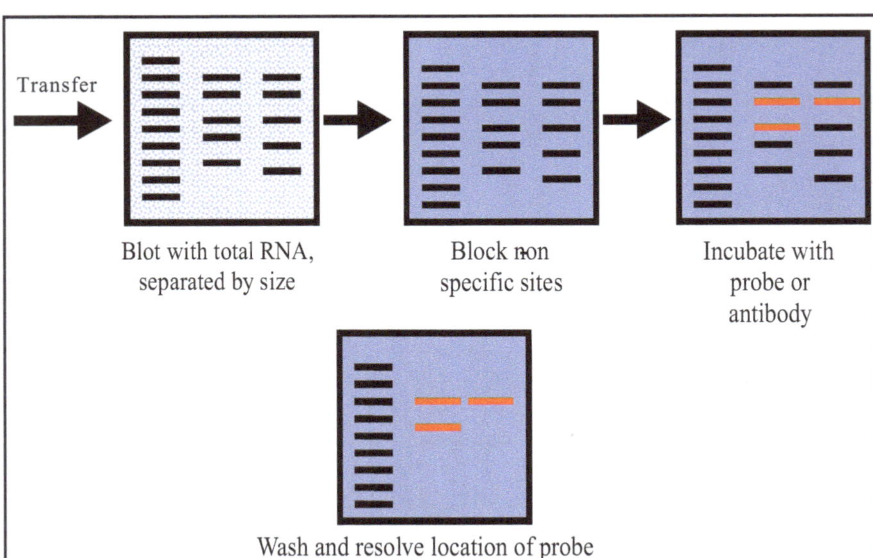

Principle

The manifestation of gene expression in cellular processes is by its transcription into mRNA, which subsequently is translated into a protein. Thus, the ability to isolate clean, intact mRNA is vital in order to identify expressed genes and, therefore, essential in the analysis of gene expression in transgenic plants. In order to elucidate the regulatory properties of a gene, it is necessary to know the sequence and level of mRNA produced from that gene. Northern analysis involves the separation of RNA on an agarose gel and its transfer to a nylon membrane. The size and amount of specific mRNA is determined by hybridizing a specific labelled probe to the nylon filter, allowing the determination of size of the entire message and also the level of the message. Northern analysis is the most commonly used technique to detect cell and tissue specific gene expression and to detect the expression of the gene introduced into transgenic plant cells (Burnette 1981).

Reagents

GITC buffer (200 ml)

94.53 g Guanidine isothiocyanate (4 M final)

1.67 ml 3 M NaOAc, pH 6

Bring volume to 200 ml

Add 1.67 ml ß-mercaptoethanol

Store at 4°C and away from light

CsCl buffer (100 ml)

95.97 g CsCl (5.7 M final)

0.83 ml 3 M NaOAc, pH 6

Bring volume to 100 ml

Sterile filter with 0.8 µm

(This is optional).

10X MOPS

0.4 M 3-[N-Morpholino] propanesulfonic acid, pH 7.0

0.1 M Sodium acetate

10 mM EDTA (pH 8.0)

Prepare in DEPC water. OK if solution turns yellow.

Sterilize by autoclaving

DEPC water

Add DEPC to ddH_2O at 1:1000 dilution

Let it sit overnight at room temp (destroys RNAase)

Autoclave the next day (destroys DEPC)

RNA sample buffer (7.25 ml)

0.5 ml 10X MOPS

1.75 ml 37% Formaldehyde

5.0 ml Formamide

Keep away from light!

Store at -20°C

RNA loading buffer

50% Glycerol

1 mM EDTA

0.25% Bromophenol blue

0.25% Xylene cyanol

Use DEPC water, keep at 4°C

2X Slaman hybe

2% SDS

2 M NaCl

Will not be in solution at RT

Keep at 4°C

20X SSPE

3.6 M NaCl

0.2 M NaH_2PO_4-H_2O

0.02 M EDTA

pH to 7.7 1

Protocol

Preparation

- ❐ Everything must be thoroughly cleaned of RNAase, and gloves must be worn at all times
- ❐ Use RNAase-ZAP or some other commercially available RNAase cleaner to clean the work area and all equipment that may come into contact, directly or indirectly, with the RNA. If none are available, use a dilute solution of bleach
- ❐ Clean everything with 95 % ethanol. This will also help any residual RNAase-ZAP evaporate
- ❐ Put sign(s) around the work area that stresses the importance of a lack of contamination
- ❐ Also, use only DEPC-treated (diethylpyrocarbonate) chemicals
- ❐ Chemicals can be DEPC-treated by adding 2 ml of DEPC per litre of solution, followed by autoclaving
- ❐ Pour the formaldehyde Gel (for a 1 % gel)
- ❐ Put 0.50 g agarose into an Erlenmeyer flask
- ❐ Add 36 mL of DEPC-H_2O
- ❐ Add-5 mL of DEPC-10x MOPS
- ❐ Add 9 mL of 12.3 M (37%) formaldehyde (does not need DEPC-treatment)
- ❐ Boil in microwave to dissolve agarose. BE CAREFUL to inhale as little of the formaldehyde fumes as possible
- ❐ Pour gel into set up box with comb in it. If using 66 p-L loading samples as described here, 3 wells must be taped

together to form I using the 8-welled comb in order for the sample to fit. For any sized loading, - sample, DO NOT use the outer wells

Loading the Gel (66 ml sample described, but ratios are what is important)

- ❑ Dilute 30 mg of quantified tRNA to 12.1 ml in DEPC-H_2O
- ❑ Add 5.5 ml of 10x MOPS
- ❑ Add 9.9 ml of 12.3 M formaldehyde
- ❑ Add 27.5 ml of formamide
- ❑ Incubate in a PCR machine for 15 minutes at 55° C program 16 (I x cycle 3 1) in PCR machine
- ❑ Add 11 ml of formaldehyde running buffer (RNA loading dye)
- ❑ Add to appropriate well, avoiding outer wells

Running the Gel

- ❑ Perform electrophoresis at 100 volts for 1 hour or until loading dye has run 2/3 to ¾ of the way through the gel
- ❑ If you do stain, use ethidium bromide in DEPC-0.5 M ammonium acetate
- ❑ For destaining, use DEPC-0.5 M ammonium acetate
- ❑ Before moving gel, let gel cool a little or it may break because formaldehyde, gels are much weaker than normal agarose gels. Even after cooling, be exceedingly careful

Vacuum Blotting

- ❑ Rinse the gel in a dish with DEPC-H_2O
- ❑ While rinsing set up the vacuum blotter (clean it for RNAase first) (27-37)

- Attach a water trap (side-arm flask) to the vacuum blotter with a tube
- Attach the water trap to the provided regulator port (thing with the pressure gauge) with a tube
- Attach the regulator port to the vacuum (huge object on wheels [actually a freeze-drier]) with a tube
- Cut a piece of Whatman 3mm paper and a piece of Hybond N^+ nylon membrane paper on one of the gel patterns. Be especially careful to keep these RNAase-free
- Wet the membrane in DEPC-H_2O
- Wet the membrane and the filter paper in DEPC-10X SSC
- Put the porous vacuum plate on the main stage of the vacuum blotter
- Put the wetted filter paper on the porous vacuum plate in the center with long ends of each rectangle parallel to each other
- Put the wetted nylon membrane on top of the filter paper
- Wet the seal o-ring (on the main stage of the vacuum blotter) with DEPC-H_2O
- Put the window gasket that has been precut for the appropriate sized gel on next- The window gasket must overlap the nylon membrane by at least 5 mm, and it should be consistent on all sides. You will probably have to adjust the positioning of the filter paper and the nylon membrane. The window gasket must also cover the entire reservoir seal o-ring to ensure a vacuum
- Place the gel on the nylon membrane such that the edges are overlapped by the window gasket. Be certain that all of the area containing RNA is touching the nylon membrane, but do not move the gel once it touches the nylon membrane
- Place the sealing frame over the entire setup
- Unscrew the bleeder valve on the vacuum regulator

- ❐ Hold the gel in place with your gloved hands, while another person turns on the vacuum
- ❐ Have them adjust the bleeder until the vacuum regulator reads a pressure of 5 inches Hg, while you hold the gel in place
- ❐ While holding the gel in place, add 1 litre of DEPC-10X SSC to the transfer reservoir so that the gel is completely covered by the transfer solution
- ❐ Put the lid on
- ❐ Allow transfer to occur at 5 inches Hg, for 1.5-2.5 hours. Effective transfer has not been established outside this pressure or time of transfer range, but the instruction manual contains information suggesting deviation from this protocol is detrimental to effective transfer.

Preparation for Probing

- ❐ Disconnect the vacuum from the vacuum blotter, and turn the vacuum off
- ❐ Remove the sealing frame and allow the transfer solution to drain off
- ❐ Remove the gel
- ❐ If you wish to estimate transfer efficiency and determine whether RNA was present in the gel to begin with (all of it will not have transferred), you may stain it with ethidium bromide in DEPC-0.5 M acetate and destain it in DEPC 0.5 ammonium acetate
- ❐ Remove the window gasket, and nylon membrane
- ❐ Soak the membrane in DEPC-2x SSC for 5 minutes or just rinse in DEPC-2x SSC
- ❐ Wrap the blot in Saran Wrap
- ❐ UV-crosslink both sides of the blot in the UV-crosslinker

- Rinse all of the blotting equipment with DEPC-H_20 and dry it. Store all the equipment in an out-of-the-way place not likely to be disturbed by RNAase-laden people. Put these things in plastic if possible.

Probing

- For the most part, the Northern blot can be probed as normal
- In order to be certain of transfer, each northern blot should first be probed with a probe of the "gene" encoding 18s rRNA. These probes will work for at least 1 month
- Perform all washes as normal and put on film for about 1 hour for a fresh probe and proportionately less for older probes
- After developing, trace the position of the northern blot onto the developed film, so that you will know where the ribosomal bands are for future probes. There should be 1 visible band. The darkness of these bands can be used a loading control
- Stop the northern blot with boiling 0.5 % SDS, and allow it to cool to room temperature
- Put it on film for 3 days to ascertain the efficiency of stripping
- Probe with a probe of the desired gene as normal
- Only perform the 1st wash 2x at room temperature
- Put on film for 1 day to 1 week depending on circumstances. Generally, 3 days is optimal
- Develop and trace the position of the northern blot onto the film, and compare it with the film from when the northern blot was probed with the ribosomal RNA probe. Some nonspecific binding will probably occur at the location of the ribosomal band, so ignore that and look for other bands. There will also probably be a band toward the lower end of mRNA length. It is not yet known what this band corresponds to, but it has been present every time northern blots have been probed with cDNA probes

- The northern blot may be probed with other probes after stripping it, but after stripping it 4 times, there is a significant loss of efficiency of probe binding, probably because some of the RNA is removed with each time it is stripped.

Applications

Gene expression

Northern blotting technique is widely used to find gene expression and regulation of specific genes. For example if we find human insulin like gene in oyster, then by isolating and immobilizing RNA and blotting it with insulin probe we can tell whether the gene is expressing or not.

Disease Identification

By using blotting technique we can identify infectious agents present in the sample and inherited disease can be identified.

Mapping

It can be applied to mapping restriction sites in single copy gene.

Troubleshooting

Poor signal

- Probe specific activity too low
- Inadequate transfer buffer
- Final wash too stringent
- Hybridization time too short
- Inappropriate membrane

Northern blot transfer problem

Change the concentration of formaldehyde gels

Northern blot problem

If the probe is dirty, there may be unincorporated radionuclide in the mixture which will give high background and obscure the signal from the probe.

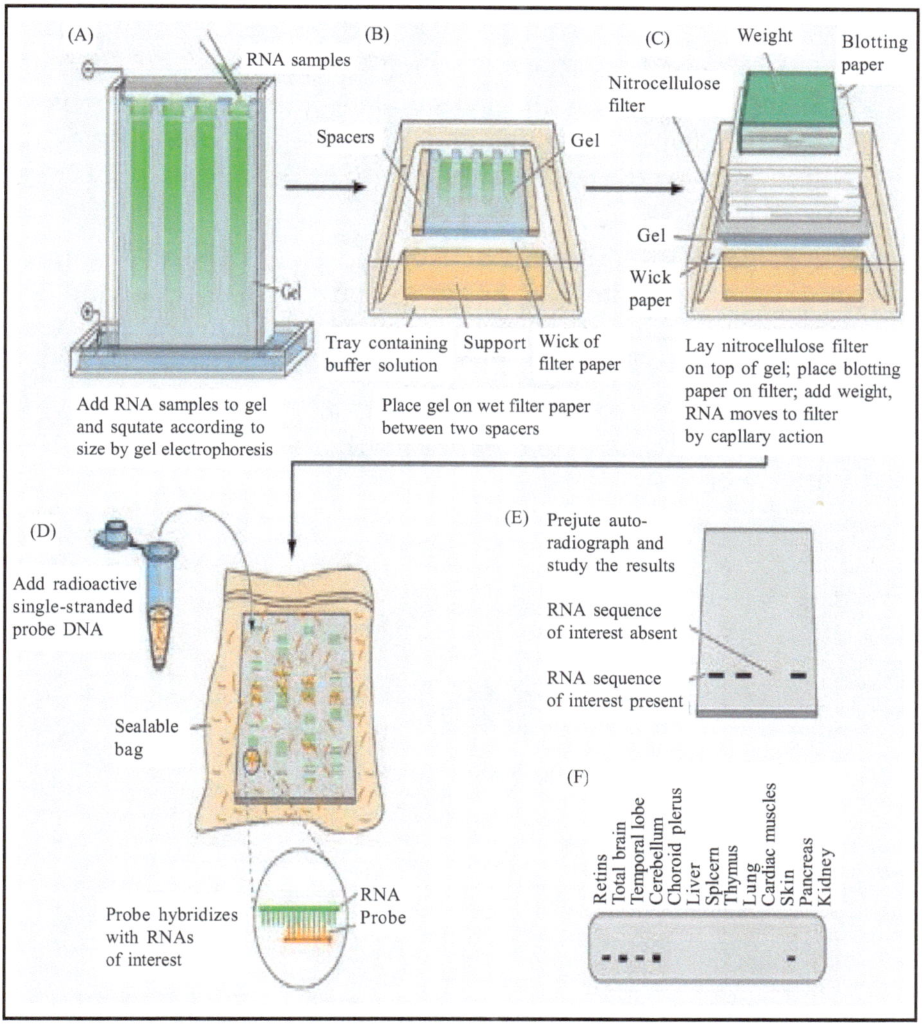

Figure 9: Developmental Northern blotting. (A–E) Procedure for Northern blotting. (A) RNA is isolated from various tissues and is separated by size using gel electrophoresis. (B) The gel is then placed on a paper wick, which absorbs an ionic solution from a trough. (C) A filter that traps RNA is placed above the gel, and blotting paper is placed above the filter. Capillary action draws the solution through the gel, trapping the RNA on the filter. (D) The filter is incubated with radioactive single-stranded DNA complementary to the mRNA of interest. (E) After any unbound DNA is washed off, autoradiography localizes the mRNA in the samples that contain it. (F) Drawing of a developmental Northern blot showing the presence of Pax6 mRNA in the eye, brain, and pancreas of a mammalian embryo.

References

Burnette W. N. (1981) Western blotting: electrophoretic transfer of proteins from sodium dodecyl sulfate – polyacrylamide gels to unmodified nitrocellulose and radiographic detection with antibody and radioiodinated protein A. *Anal Biochem* 112 (2): 195-203.

Jackson, J. (2002). Detection of Nucleic Acids Using Chemiluminescence: From Northerns to Southerns and Beyond, *In Luminescence Biotechnology: Instruments and Applications*, Van Dyke, K., *et al.*, Ed., CRC Press, pp. 223-230.

Sambrook, J., Fritsch, E.F., & Maniatis, T. (1989) *In Molecular Cloning: A Laboratory Manual*. Cold Spring Harbor Laboratory Press, NY, Vol. 1, 2, 3.

□□□

Chapter 10

Cloning

Introduction

Only after the discovery of restriction endonucleases it became possible to digest any DNA molecule at a specific site and any two pieces of DNA could be joined together with the help of another modifying enzyme, the DNA ligase. Consequently, a new branch of molecular biology, namely genetic engineering has been developed. With the help of genetic engineering techniques, it is now possible to identify and isolate the gene (for a protein of interest and insert it into the genome of another organism with the help of a suitable vehicle. The foreign gene thus cloned, can be amplified to produce millions of its copies. The insert gene can be expressed by the host to synthesize the gene product. The technique can also be used for understanding the regulation of expression of the cloned gene. The possibilities are endless and probably sky is the only limit. In this chapter we will briefly discuss the basic principles and protocol involved in gene cloning.

Principle

Principle behind cloning can be compared with the work of a tailor, who cuts a clot in desired manner and joins many pieces to design a dress. Using this technology, referred as the recombinant DNA (or rDNA) technology, it is possible to identify the pieces of DNA which code the protein of interest, isolate these pieces and clone them into the genome of a totally unrelated organism. This has provided a very powerful tool in the hands of the molecular biologist which has enormous applications.

Why to Clone DNA?

- A particular gene can be isolated and its nucleotide sequence determined
- Control sequences of DNA can be identified & analyzed
- Protein/enzyme/RNA function can be investigated
- Mutations can be identified, e.g. gene defects related to specific diseases
- Organisms can be 'engineered' for specific purposes, e.g. insulin production, insect resistance, etc.

Basic Steps of Gene Cloning

- A fragment of DNA , containing the gene to be cloned, is inserted into a circular DNA molecule (vector) "Recombinant DNA molecule" or "Chimera"
- The vector acts as a vehicle that transports the gene into a host cell (usually, bacterium) possibly other types of living cell
- Within the host cell the vector multiplies, producing numerous identical copies not only of itself but also of the gene that it carries
- When the host cell divides, copies of the recombinant DNA molecule are passed to the progeny and further vector replication takes place
- After a large number of cell divisions, a colony, or clone, of identical host cells is produced. Each cell in the clone contains one or more copies of the recombinant DNA molecule

There are many types of cloning and new inventions are continuous but in this chapter we have explained Cell-based DNA cloning in vice.

Cell-based DNA cloning

DNA can be inserted into a cell by:

- Electroporation
- Transformation
- Protoplast fusion
- Microinjection
- Gene gun

Agro bacterium-mediated Transformation of Rice (Raineri *et al.*, 1990)

Reagents

Sterilization of seed material

The seed coat was removed and the seeds treated with 500 ppm. Streptomycin for 1 h. Seeds were washed with sterile water once, treated with 1% Bavistin for 15 min and then washed 5 times with sterile water. Following this, the seeds were treated with 4% sodium hypochlorite for 15 min and washed once with sterile water. Finally seeds were immersed in a 0.1% solution of $HgC1_2$ for 5 min and then washed 5 times with sterile water to remove the residue of $HgC1_2$.

Preparation of rice explants for co-cultivation

Seeds were placed on MS medium with optimal hormone concentrations for the promotion of callus induction. The Petri plates were incubated at 25° C in growth chamber under light for 4 days. Radicles and plumules were removed and endosperm was cut from scutellum. Scutellum was added to induction medium and incubated at 28° C in the dark for 72 h. Calli of one generation were used as explants for shooting.

Preparation of culture for co-cultivation

A single *Agrobacterium* colony was taken from YEP (Yeast extract, 10 g/l; peptone, 10 g/l; sodium chloride, 5 g/l; pH 7.0) plate containing *Agrobacterium* culture. It was inoculated into a conical flask containing 20 ml YEP broth and 50 mg/l Kanamycin. It was incubated overnight at 28°C in an orbital shaking incubator. The culture was used for transformation when it reached an OD of 0.9-1.0 at 600 nm. The culture was then harvested by centrifugation (5,500 rpm for 10 min at 4°C).

Protocol for co-cultivation

Rice explants (calli/scutellum) ready for inoculation were mixed with *Agrobacterium* suspension. Inoculated explants were blotted dry with sterile filter paper to remove excess bacteria. Then calli were transferred onto a piece of filter paper placed onto a co-culture medium in Petri dishes sealed with Parafilm. Co-cultivation was carried out in the dark at 28°C for 3 days. After 3 days of co-cultivation explants were washed with *Yz* MS-Cef medium, blotted dry and then transferred to MS-Cef-Km medium.

Using scutellum as explant material for co-cultivation

- Scutellum and *Agrobacterium* cells were mixed with constant stirring for co-cultivation in liquid co-culture medium.
- Co-cultured scutella were transferred to the co-culture medium
- Co-cultivation was carried out in the dark at 28^0 C for 3 days
- Co-cultivated scutella were washed with sterile water twice and then washed with half MS-Cef medium.
- Scutella were placed on callus induction medium MS-D-Cef-km medium and sub-cultured onto the same medium.
- Calli were transferred to the regeneration medium. Regenerated plants were transferred to RE2 medium for full plant formation. After profusion, rooting plants were hardened.

NOTE: *Use 60 μM Acetosyringone at all 3 stages of co-cultivation.*

Confirmation of integration of GUS (Escherichia coli P-D-glUCUronidase gene) *gene*

GUS Histochemical Assay

Histochemical staining of the GUS gene was performed to confirm the presence of transgene. The substrate used for localization of J3-glucuronidase activity in cells is 5- bromo-4-chloro—3-indolyl glucuronide (X-Gluc) which gives a blue precipitate at the site of enzymatic activity .The product of J3-glucuronidase action is colourless. The derivative produced must undergo an oxidative dimerization to form the insoluble and highly coloured blue dye.

Staining solution

50 mM sodium phosphate buffer (pH 7)

1 mM X-Gluc

0.1% v/v Triton

4 mM potassium ferrocyanide

100 µg/ml chloramphenicol

Protocol

- Proliferated callus obtained from the putatively transformed colony were immersed in the GUS staining solution and incubated overnight at 37 °C. The following day, the staining solution was added and kept at room temperature to remove the pigmentation.
- GUS expression cells were detected microscopically by a distinct blue coloration due to enzymatic cleavage of 5-bromo-4-chloro-3-indolyl glucuronide.

Media used for transformation of rice varieties

S. No.	Medium	Composition
1	MS-D (Bottom medium)	MS basal, 2mg 2,4-D+1mg Kinetin, 8g/1 agar
2	MS-D-Km	MS-D, 50ppm Kanamycin
3	RE1	MS basal, 4mg Kinetin +1mg IAA, 30g Mannitol, 2g Casaminoacid, 8g/1Agar. pH 5.8
4	RE1-Km	RE1, 50ppm Kanamycin
5	RE2	MS basal, 30g Mannitol, 2g Casaminoacid, 8g/1 Agar. pH 5.8
6	RE2-Km	RE2, 50 ppm Kanamycin

Biolistic/Gene Gun transformation protocol for Rice (Wang *et al.*, 1988)

Preparation of microprojectiles

1. Weigh 60 mg tungsten powder into a 1.5 ml microfuge tube. This is sufficient for 100 shots.
2. Add 1 ml of absolute ethanol -use a freshly opened bottle.
3. Centrifuge for 10 sec in microfuge and carefully remove ethanol with a pipette.
4. Add 1 ml water, resuspend tungsten and centrifuge. Repeat 3 times.
5. Finally resuspend the tungsten particles in 1.0 ml of 50% glycerol.
6. Store at -20° C and use within 2 months of preparation.

Coating of plasmid DNA to microprojectile

1. Vortex the microprojectile suspension prepared in 50% glycerol (60 mg/ml) for at least 5 min on a platform vortex to resuspend and disrupt agglomerated particles.

2. Pipette 5μl of microprojectile suspension into a 1.5 ml microfuge tube. It is important to vortex the tube containing the microprojectile suspension continuously in order to maximize uniform sampling.
3. Whilst vortexing vigorously, add in the following order:
   - 5μ1 DNA (μg/ml)
   - 50μl 2.5M $CaCl_2$
   - 20μ1 spennidine (free base, 0.1 M)
4. Continue vortexing for 3 min.
5. Allow the microprojectiles to settle for 3 min.
6. Pellet microprojectiles by spinning 2 sec in microfuge (approx 5,000 rpm).
7. Remove liquid and discard.
8. Add 150 μl of 70% ethanol without disturbing the pellet.
9. Remove liquid and discard.
10. Add 150 μl 100% ethanol (99.5% ethanol can also be used).
11. Remove liquid and discard.
12. Add 55-60 μ1 100% ethanol (99.5% ethanol also can be used).
13. Resuspend the pellet by tapping the side of the tube
14. Remove 6 μl aliquots of microprojectiles and transfer them to the centre of the macrocarrier.

Preparation of callus for shooting

Approximately 10-12 calli, each about 0.5 cm diameter, were sliced into 2-3 mm pieces and placed in the middle of a petri plate containing osmotic medium (MS basal medium, 0.5 M. D-Mannitol, 30g/1 sucrose, 2.5 mg /1 2,4 D and 3g/1 Phytagel (Vain *et al.*, 1993) for 4 h prior to bombardment. The callus tissue was left on the medium overnight following the bombardment before being transferred to subculture medium.

Shooting of explant

For each shot you will need 1100 psi rupture disc (the small dark brown discs), one flying disc (the large orange discs), and one stopping screen (the wire mesh screens). All these are supplied in the BioRad shooting kit. Before shooting, sterilize the number of discs and stopping screens that you will need by placing them for 5-10 min in a Petri plate containing absolute ethanol. The discs and screens are dried by standing them up along the side of a sterile Petri plate and allowing them to air-dry, under sterile conditions in the tissue culture hood.

While sterilizing the disposable shooting components, the gun parts should also be surface sterilized by wiping with absolute ethanol. The components can be removed from the gun vacuum chamber for surface sterilization. These include the holder for the rupture discs, the stage and holder assembly for the flying discs, and the platform for the leaf samples. After removing these and wiping with ethanol, the interior of the vacuum chamber and door should be wiped with absolute alcohol. Finally the five metal ring holders for the flying discs should also be surface sterilized.

GUS Histochemical Assay

Histochemical staining of the GUS gene was performed to confirm the presence of transgene. The substrate used for localization of J3-glucuronidase activity in cells is 5- bromo-4-chloro – 3-indolyl glucuronide (X-Gluc) which gives a blue precipitate at the site of enzymatic activity .The product of J3-glucuronidase action is colourless. The derivative produced must undergo an oxidative dimerization to form the insoluble and highly coloured blue dye.

Staining solution

50 mM sodium phosphate buffer (pH 7)

1 mM X-Gluc

0.1% v/v Triton

4 mM potassium ferrocyanide

100 µg/ml chloramphenicol

Procedure

- Proliferated callus obtained from the putatively transformed colony were immersed in the GUS staining solution and incubated overnight at 37^0 C. The following day, the staining solution was added and kept at room temperature to remove the pigmentation.
- GUS expression cells were detected microscopically by a distinct blue coloration due to enzymatic cleavage of 5-bromo-4-chloro-3-indolyl glucuronide.

Applications

Agricultural applications

- The technique can be used to understand the regulation of expression of particular gene in a non-ambiguous manner
- To understand the structure-function relationship of biomolecule
- To see the mutual co-ordination between different genes
- To produce a biomolecule by the cell of one's choice or by an organism, which does not produce the molecule under normal physiological conditions.
- They have proved to be extremely valuable tools in studies on plant molecular biology, regulation of gene action, identification of regulatory/promotary sequences, etc.
- Specific genes have been transferred into plants to improve their agronomic and other features.
- Genes for resistance to various biotic stresses have been engineered to generate transgenic plants resistant to insects, viruses, etc.
- Several gene transfers have been aimed at improving the produce quality.
- Transgenic plants are being used to produce novel biochemicals like hirudin, etc. which are not produced by normal plants.

- Transgenic plants can be used vaccines for immunization against pathogens; this is fast emerging as an important objective.

Troubleshooting

Un-induced colonies

Growth of un-induced colonies on selective plates suggests incomplete digestion of the gap-repair construct during linearization. However, the number of colonies should be low. Otherwise repeat with improved digestion and/or gel purification of the linearised gap repair construct.

References

Angrand, P.O., Daigle, N., van der Hoeven, F., Scholer, H.R. & Stewart, A.F. (1999) Simplified generation of targeting constructs using ET recombination. *Nucleic Acids Res.* 27.

Copeland, N.G., Jenkins, N.A. & Court, D.L. (2001) Recombineering: a powerful new tool for mouse functional genomics. *Nat Rev Genet.* 2, 769-779.

Hall, R.D., Riksen-Bruinsma, T., Weyens, G.J., Rosquin, I.J., Denys, P.N., Evans, I.J., Lathouwers, J.E., Lefebvre, M.P., Dunwell, J.M., van Tunen, A. & Krens, F.A. (1996) A high efficiency technique for the generation of transgenic sugar beets from stomatal guard cells. *Nat Biotechnol.*14, 1133–1138.

Lee, E.C., Yu, D., Martinez de Velasco, J., Tessarollo, L., Swing, D.A., Court, D.L., Jenkins, N.A. & Copeland, N.G. (2001) A highly efficient Escherichia coli-based chromosome engineering system adapted for recombinogenic targeting and sub cloning of BAC DNA. *Genomics.*73, 56–65.

Raineri, D., Bottino, M. P., Gordon M. P. & Nester E. W. (1990) *Agrobacterium*-Mediated Transformation of Rice (*Oryza sativa* L.) *Nature Biotechnology* 8, 33 - 38.

Takken, F.L., Van, W.R., Michielse, C.B., Houterman, P.M., Ram, A.F. & Cornelissen, B.J. (2004) A one-step method to convert vectors into binary vectors suited for Agrobacterium-mediated transformation. *Curr Genet.* 45, 242–248.

Wang, Y.C., Klein, T.M., Fromm, M.E., Cao, J., Sanford, J.C. & Wu, R. 1988. Transformation of rice, wheat, and soybean by the particle bombardment method. *Plant Molec Biol* 11, 433-439.

Yu, D., Ellis, H.M., Lee, E.C., Jenkins, N.A., Copeland, N.G. & Court, D.L. (2000) An efficient recombination system for chromosome engineering in Escherichia coli. *Proc Natl Acad Sci* USA. 97, 5978–5983.

Yanisch-Perron, C., Viera, J. & Messing, J. (1985) "Improved M13 phage cloning vectors and host strains: nucleotide sequences of the m13mp18 and pUC19 vectors". *Gene.* 33, 103-199.

Zhang, Y., Buchholz, F., Muyrers, J.P. & Stewart, A.F. (1998) A new logic for DNA engineering using recombination in Escherichia coli. *Nat Genet.*20, 123–128.

❒❒❒

Chapter 11

Sequencing of Nucleic Acids

Introduction

DNA sequencing is the process of determining the nucleotide order of a given DNA fragment. The most commonly used sequencing method is the dideoxy method. This method uses dideoxynucleotide triphosphates (ddNTPs) which have an H on the 3′ carbon of the ribose sugar instead of the normal OH found in deoxynucleotide triphosphates (dNTPs). Dideoxynucleotides are chain terminators. In a synthesis reaction, if a dideoxynucleotide is added instead of the normal deoxynucleotide, the synthesis stops at that point because the 3′OH necessary for the addition of the next nucleotide is absent (Sanger *et al.*, 1977).

Principle

The purpose of sequencing is to determine the order of the nucleotides of a gene/organism. DNA sequencing reactions are not started from genomic DNA (like in PCR) but mostly from PCR fragments or cloned genes.

The sequencing reaction

There are three major steps in a sequencing reaction (like in PCR), which are repeated for 30 or 40 cycles.

Denaturation at 94°C: During the denaturation, the double strand melts open to single stranded DNA, all enzymatic reactions stop (for example: the extension from a previous cycle).

Annealing at 50°C: In sequencing reactions, only one primer is used, so there is only one strand copied (in normal PCR: two primers are used, so two strands are copied). The primer is jiggling around, caused by the Brownian motion. Ionic bonds are constantly formed and broken between the single stranded primer and the single stranded template. The more stable bonds last a little bit longer (primers that fit exactly) and on that little piece of double stranded DNA (template and primer); the polymerase can attach and starts copying the template. Once there are a few bases built in, the ionic bond is so strong between the template and the primer, that it does not break anymore.

Extension at 60°C: This is the ideal working temperature for the polymerase (normally it is 72°C, but because it has to incorporate dNTPs which are chemically modified with a fluorescent label, the temperature is lowered so it has time to incorporate the 'strange' molecules. The primers, where there are a few bases built in, already have a stronger ionic attraction to the template than the forces breaking these attractions. Primers that are on positions with no exact match, come loose again and don't give an extension of the fragment.

The bases (complementary to the template) are coupled to the primer on the 3'side (adding dNTP's from 5' to 3', reading from the template from 3' to 5' side, bases are added complementary to the template). When a dNTP is incorporated, the extension reaction stops because a dNTP contains an H-atom on the 3rd carbon atom (dNTP's contain a OH-atom on that position). Since the dNTP's are fluorescently labeled, it is possible to detect the color of the last base of this fragment on an automated sequencer.

Protocol

Method: Sequencing of double stranded DNA using dideoxy chain termination (Sanger method DNA sequencing)

Principle

This method is used to sequence double stranded DNA, such as a plasmid insert or purified PCR product. The dideoxy chain termination (or enzymatic) method of DNA sequencing involves the in vitro synthesis of a DNA strand by a modified bacteriophage T7 DNA polymerase (SequenaseR/USB) using a single stranded DNA template. Synthesis is initiated at a specific site on the sequencing template, as determined by the annealed sequencing oligonucleotide primer. The synthesis reaction is terminated by the incorporation of a dideoxynucleotide analog (dNTP) that will not accept further elongation of the synthesized strand. Four separate reactions are set up for each template to be sequenced, each containing one nucleotide as the dNTP plus all four deoxynucleotide (dNTPs). For each reaction, enzyme catalyzed synthesis will be terminated in a fraction of the population of chains at each site where the dNTP is incorporated (e.g. in the dGTP reaction tube, a population of chains result with a fraction of the chains terminated at each site a "G" occurs). Radio labeled dATP is included in the synthesis, and the products are size separated by high resolution denaturing polyacrylamide gel electrophoresis and visualized by autoradiography. The Sequenase kit also includes the reagents for sequencing reactions using dITP (in place of dGTP). This set of reactions is used only when you wish to attempt to resolve a problem with a compression (i.e. when the secondary structure of the DNA interferes with the size separation in the sequencing gel).

Time required

2-4 Days

- Pouring of sequencing gel: 1-2 hours, polymerization 1 hour to overnight
- Sequencing reactions: 1 hour and 30 minutes (with experience)

- High-resolution denaturing polyacrylamide gel electrophoresis : 4-8 hours
- Removal of Urea and Gel Drying: 1 hour 30 minutes
- Autoradiography: overnight to 2 days

Special reagents

All of the following required reagents are included in the SequenaseR kit (USB):

- Sequenase 5X buffer (200 mM Tris-HCl, 100 mM $MgCl_2$, 250 mM NaCl)
- Dithiothreitol (DTT) 0.1 M
- Labeling Mix
- Termination Mixes (for ddGTP, ddATP, ddTTP, ddCTP)
- SequenaseR enzyme
- Enzyme dilution buffer (10 mMTris-HCl, pH 7.5; 5 mM DTT, 0.5 mg/ml BSA)
- Stop solution (95% formamide, 20 mM EDTA, 0.05% Bromophenol blue, 0.05% Xylene cyanol FF).

Special Equipment

- Microcapillary flat pipet tips
- Baserunner sequencing gel unit (IBI)
- High voltage power supply

  Gel Dryer

Description of Reactions

It is important to understand the reactions occurring at each step in order to perform several sequencing reactions simultaneously because the time interval allowed for each step is critical for certain steps. The first step is the labeling step which incorporates the 32P-dATP. In this step the primer is extended using limited concentrations of dNTPs including 32P-dATP. This reaction should be allowed to continue at 4^0C for 3-5 minutes only, to result in labeled chains limited in length from several nucleotides to hundreds of nucleotides. In the second step, higher concentrations of 3 of the dNTPs plus the remaining nucleotide triphosphate as a dideoxy derivative results in termination of each of the growing chains at the next site where the dNTP can be incorporated. This reaction is carried out at 37°C and can continue for 5 minutes or longer.

Procedure

The sequencing protocol is accomplished by pouring the polyacrylamide gel on Day 1, allowing the gel to polymerize and the gel is stored overnight at room temperature wrapped in cellophane. The sequencing reactions may also be prepared on Day 1 and frozen at -20°C, then loaded on the sequencing gel the following day.

Day 1

Pouring of sequencing (8% Polyacrylamide/urea) denaturing gel

This is probably the most difficult part of sequencing, and it is recommended that taping the plates and pouring of the plug and gel be learned from an individual with experience.

Wash plates with mild soap and a non-abrasive sponge, rinse with dH_2O and clean with 100% EtOH. Siliconize only the mirrored plate with a light coat of SigmacoteR in the fume hood using a Kimwipe and a light circular motion (wear gloves) and quickly assemble plates with 0.4 mm spacers before any dust collects. Tape the bottom of the plates securely with water-proof tape and clamp sides (Do not clamp the bottom).

Recipe for 8% polyacrylamide sequencing gel

75 ml for two 45 cm plates or one 60 cm plate	**50 ml for one 45 cm plate**
31.5 g urea	21 g urea
31 ml ddH_2O	21 ml ddH_2O
7.5 ml 10X TBE	5 ml 10X TBE
15 ml 40% acrylamide/ 2% bis stock	10 ml 40% stock
375 µl 10% ammonium persulfate (APS)	250 µl 10% APS
Immediately before pouring, add:	
35µl TEMED	24 µl TEMED

The urea should be weighed out first and allowed to dissolve in the water with the aid of a magnetic stirring bar (do not use heat) while the plates are cleaned and prepared. Add the remaining ingredients, filter (0.45 micron), and leave the acryl amide solution on ice in the disposable filter container. It is advisable to de-gas the acrylamide solution by gently swirling with the vacuum on until no more bubbles are produced.

A separate plug should be made (4 ml of above acrylamide solution and 8 µl of TEMED), poured down the side of the gel and allowed to completely solidify before attempting to pour the entire gel. The TEMED should not be added to the acryl amide until you are completely ready to pour the gel. Hold the gel cast at an incline and pour the gel carefully but quickly, taking care not to trap bubbles. This is accomplished using a 60 cc syringe. It may be necessary to lightly tap the outside of the plate or alter the incline to remove bubbles. After the full length of the gel is poured, carefully insert the comb (shark's tooth or standard), place in a horizontal position and allow to polymerize at least one hour. The gel may be wrapped in cellophane and stored overnight (Longer storage is not recommended).

Sequencing Reactions (example below is for sequencing the insert in a plasmid such as pUC or Bluescript using the universal and reverse primers)

1. Prepare plasmid DNA as described elsewhere in the alkaline lysis miniprep protocol. Resuspend the pellet in 50 µl ddH_2O. Pipette an 18 µl aliquot of this into a clean eppendorf tube for sequencing.
2. Add 2 µl of 2 M NaOH/ 2mM EDTA. Incubate at room temperature for 20 minutes (alkaline denaturation).
3. Add 2 µl ammonium acetate (pH 4.5) and then 50 µl of 95% EtOH, mix well and place on ice for 15 minutes (precipitation).
4. Spin in micro centrifuge at 4°C for 10-15 minutes and carefully decant the supernatant.

5. Wash pellet with 70% EtOH, respin, and decant supernatant. Invert the tube on a Kimwipe to air dry.

The sequencing reactions may now be carried out using the SequenaseR kit and a modification of the double stranded sequencing protocol:

The 4 reaction tubes (G, A, T, C) should be labeled and the appropriate termination mix (3.5 μl) dispensed in each before beginning the protocol. Be sure to choose the dGTP labeling mix unless you intend to use the dITP reactions for resolving compressions.

1. Resuspend the pellet in
   - ❑ 6 μl ddH_2O
   - ❑ 2 μl 5X Sequencing buffer
   - ❑ 2 μl Sequencing primer (0.5 pmol/μl)
   - ❑ or, 2 μl 5X Sequencing buffer, 0.5-1.0 pmol primer, and the appropriate amount of water to a total volume of 10 μl
2. Incubate at 37°C for 15 minutes for annealing of primer. (Alternatively, the tube may be heated to 65°C for 2 minutes and allowed to cool to room temperature over 30 minutes. This has not been found to be necessary in my experience if the alkaline denaturation step is used)
3. Add 2 μl diluted labeling mix 1:5 to 1:15. (Note: for sequencing within 30 bases of primer, dilution should be 15 fold)
   - ❑ 1 μl 100 mM DTT
   - ❑ 1 μl 32P-dATP
   - ❑ 2 μl diluted Sequenase (diluted 1:8 in ice-cold TE immediately before use)

   Allow reaction to extend at room temperature (or at 4°C) for 3-5 minutes.
4. Dispense 3.5 μl of the above mixture into each of 4 prelabeled tubes (G, A, T, C) containing 2.5 μl of appropriate termination mix (ddGTP, ddATP, ddTTP, ddCTP) and incubate at 37°C for 5 minutes or longer

5. Add 4 μl of Sequenase stop solution to each tube. Samples may be stored at -20°C until loaded on the sequencing gel

Day 2

Remove the sequencing gel from refrigerator, place on Baserunner, and attach buffer chambers. After filling the buffer chambers with 1X TBE, gently flush out the wells with a syringe and needle and pre run for 30 minutes to 1 hour.

1. Thaw the sequencing reaction samples, heat to 75-80°C for 2 minutes and immediately load 1-3 μl from per lane. Use the specialized microcapillary flat pipette tips to load the gel. Slowly layer each sample into the bottom of each well, avoiding bubbles.
2. The lanes should be loaded G, A, T, C in adjacent order for each sequencing reaction. Run at a *constant power of 55 watts,* with voltage from 1800 to 2000 V or greater. A second aliquot of the samples may be heated again and reloaded after a variable period of time in order to read further. Ideally, an area of overlap between the bottom of the first load (larger fragments which have run longer) and the top of the second load (smaller fragments which have run less time) will be found on the autoradiogram. (One recommendation is to run the first load twice the amount of time it takes the bromophenol blue to run off, and the second load 2-2.5 hrs).
3. At the completion of the run, turn off power supply, carefully drain buffer chambers (buffer must go in liquid radioactive waste container), and remove the gel from the apparatus.
4. Place the gel with the glass plate face down on a flat surface. Remove the mirrored glass plate (the gel should stick to the unsiliconized glass plate). Leave the spacers in place on each side.
5. Carefully soak the gel for 10-15 minutes in 10% glacial acetic acid on the glass plate with minimal rocking (be careful that the gel doesn't float off the glass plate).

6. Remove the gel from the acetic acid. Cut two pieces of Whatman paper the size of the glass plate and carefully place on top of the gel with glass plate below. Turn the Whatman paper, gel and glass plate over as a unit and then remove the glass plate (the gel sticks to the Whatman). Cover the gel with Saran wrap and dry in a gel dryer for 1-1.5 hours.
7. Expose film overnight (usually sufficient for 32_P (Potassium isotope 32)) or longer.

Reagents

- Sigmacote
- Acrylamide
- N,N′-Methylenebisacrylamide (electrophoresis grade)
- Ammonium persulfate
- TEMED (N,N,N′N′ Tetramethylethylenediamine)

Recipes for Acrylamide stocks
40% Acrylamide/ 2% bis stock

Acrylamide	38 g
N,N′-methylene bisacrylamide	2 g
dH_2O	to 100 ml

Safety Considerations: Acrylamide is a neurotoxin (especially in solid form). Work with this substance should be done in the hood. Wear gloves and a mask when weighing out and handling. Clean up all spills immediately.

Applications

Whole genome analysis

The most obvious application of DNA sequencing technology is the accurate sequencing of genes and genomes. Only about 500-800 bases can be sequenced in one experiment so larger DNA molecules, including whole genomes, must be broken into smaller fragments before sequencing and then reassembled by searching for overlaps. Accuracy is achieved by sequencing each template several times.

Deriving meaningful knowledge from DNA sequence will define biological research through the coming decades and require the expertise and creativity of teams of biologists, chemists, engineers, and computational scientists, among others. A sampling follows of some research challenges in genetics.

- Gene number, exact locations, and functions
- Gene regulation
- DNA sequence organization
- Chromosomal structure and organization
- Noncoding DNA types, amount, distribution, information content, and functions
- Coordination of gene expression, protein synthesis, and post-translational events
- Interaction of proteins in complex molecular machines
- Predicted vs experimentally determined gene function
- Evolutionary conservation among organisms
- Protein conservation (structure and function)
- Proteomes (total protein content and function) in organisms
- Correlation of SNPs (single-base DNA variations among individuals) with health and disease
- Disease-susceptibility prediction based on gene sequence variation
- Genes involved in complex traits and multigene diseases
- Complex systems biology including microbial consortia useful for environmental restoration
- Developmental genetics, genomics

Troubleshooting

No sequence data

Priming site not present

- If you've chosen one of the sequencing facility's vector primers, make sure it is present in your vector.
- Double-check your plasmid maps/sequences.

Not enough or no DNA/primer in tube

- Double-check your quantitations, stock concentrations and dilutions.

Inhibitory contaminant

- The cycle sequencing reaction used to amplify samples for automated sequencing is very sensitive to the presence of certain contaminants, some of which will completely inhibit our sequencing enzyme. Please check the Contaminant chart in the Template preparation and purification

Noisy data with weak signal

- "Noisy" data can be identified by the presence of multiple peaks and numerous "N"s within your sequence. The Sequencing Analysis program assigns an "N' as a base identification when there are two or more peaks present at one position. This "N" may signify the legitimate occurrence of two nucleotides, as in the case of a heterozygote, but may also be seen when background noise is high or when multiple products are present.

Inhibitory contaminant e.g. salts, phenol

- The cycle sequencing reaction used to amplify samples for automated sequencing is very sensitive to the presence of certain contaminants, some of which can partially or completely inhibit our sequencing enzyme.

Multiple priming sites in PCR

- This may occur when one or both of the PCR primers hybridizes to more than one position on the template DNA, giving rise to multiple PCR products. Often this will be obvious when visualizing the PCR products on an agarose gel as there will be more than one band present. In this case, gel purification of the desired product will be necessary. One can run into difficulty, however, when the products are very similar in size, which may arise when amplifying related or repetitive DNA, and do not separate well on the gel. In this case, optimization of the PCR reaction may be necessary or redesign of the PCR primers in order to choose a more specific priming site.

PCR primers acting as both forward and reverse

- Sometimes, a PCR product may be generated when one primer functions as both the forward and reverse primer in the PCR reaction, giving rise to an artifactual product. This is fairly easy to detect when sequencing the PCR product as one primer will give double peaks from the start, while the other fails to give any sequence data. Redesign your set of PCR primers.

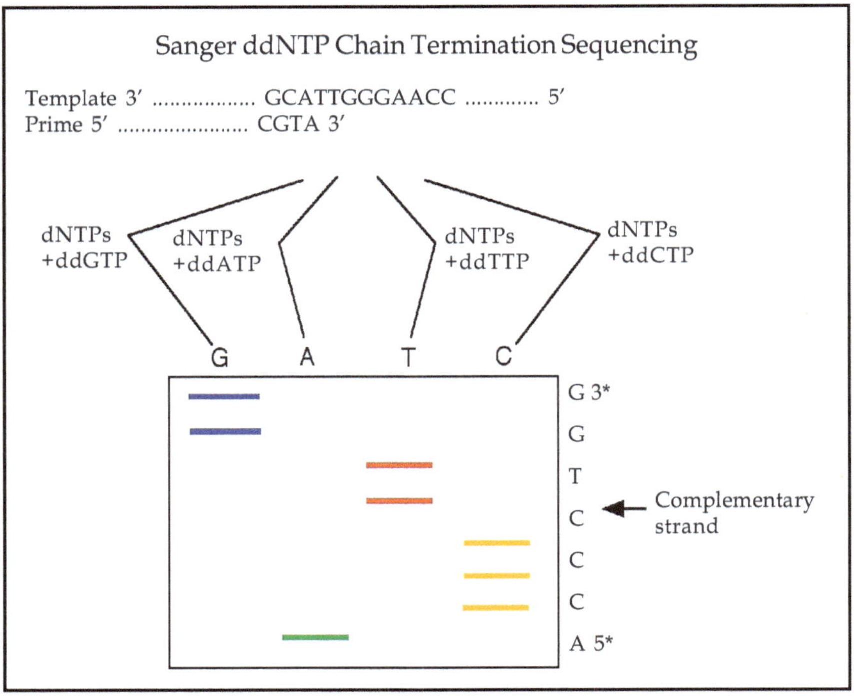

Figure 10: Schematic presentation Sanger sequencing method

Truncated sequences

- Truncated sequences can be characterized as abrupt or gradual. Abrupt truncations will show strong, clean signal up to a point and then drop sharply down over the course of a few nucleotides to much weaker or no detectable signal. Gradual truncations will show good sequence data initially but then begins to taper off to progressively weaker, smaller peaks until there is nothing but background noise. The nature of the truncation can sometimes help to determine its cause.

References

Campbell, N., Reece, J.B., (2002) Biology 6th ed., Benjamin Cummings, San Francisco.

Drlica, K., (1997) Understanding DNA & Gene Cloning, 3rd ed., John Wiley & Sons, NY.

Kreuzer, H., Massey, A., (2001) Recombinant DNA & Biotechnology, 2nd ed., ASM Press, Washington, DC.

Mardis, E.R. & Roe, B.A. (1989) Automated methods for single-stranded DNA isolation and dideoxynucleotide DNA sequencing reactions on a robotic workstation, *BioTechniques*. 7, 840-850.

Sanger, F. Niklen, S. & Coulson A. R. (1977) DNA sequencing with chain-terminating inhibitor. *Proc. Nat. Acad. Sci. USA*. 74: 5463-5467.

Sequenase, R (1989) Step by Step Protocols United States Biochemical Corporation.

Tabor, S. & Richardson C.C. (1987) DNA sequence analysis with a modified bacteriophage T7 DNA polymerase. *Proc. Nat. Acad. Sci. USA*. 84: 4767-4771.

Turner, P.C., (1997) Instant Notes in Molecular Biology, Bios, Oxford.

❑❑❑

Index

E

F

G

H

I

L

M

N

P

Q

R

S

❑❑❑